ASTEROIDS, COMETS, AND DWARF PLANETS

Greenwood Guides to the Universe
Timothy F. Slater and Lauren V. Jones, Series Editors

Astronomy and Culture
Edith W. Hetherington and Norriss S. Hetherington

The Sun
David Alexander

Inner Planets
Jennifer A. Grier and Andrew S. Rivkin

Outer Planets
Glenn F. Chaple

Asteroids, Comets, and Dwarf Planets
Andrew S. Rivkin

Stars and Galaxies
Lauren V. Jones

Cosmology and the Evolution of the Universe
Martin Ratcliffe

ASTEROIDS, COMETS, AND DWARF PLANETS

Andrew S. Rivkin

Greenwood Guides to the Universe
Timothy F. Slater and Lauren V. Jones, Series Editors

GREENWOOD PRESS
An Imprint of ABC-CLIO, LLC

Santa Barbara, California • Denver, Colorado • Oxford, England

Library of Congress Cataloging-in-Publication Data
Rivkin, Andrew S.
Asteroids, comets, and dwarf planets / Andrew S. Rivkin.
p. cm. — (Greenwood guides to the universe)
Includes bibliographical references and index.
ISBN 978-0-313-34432-9 (hard copy: alk. paper) —
ISBN 978-0-313-34433-6 (ebook)
1. Solar system. 2. Asteroids. 3. Comets. 4. Dwarf planets. I. Title.
QB501.R58 2009
523.2—dc22 2009016114

ISBN: 978-0-313-34432-9
EISBN: 978-0-313-34433-6

13 12 11 10 9 1 2 3 4 5

This book is also available on the World Wide Web as an eBook.
Visit www.abc-clio.com for details.

Greenwood Press
An Imprint of ABC-CLIO, LLC

ABC-CLIO, LLC
130 Cremona Drive, P.O. Box 1911
Santa Barbara, California 93116-1911

This book is printed on acid-free paper ∞

Manufactured in the United States of America

For my father

Contents

Series Foreword

Not since the 1960s and the Apollo space program has the subject of astronomy so readily captured our interest and imagination. In just the past few decades, a constellation of space telescopes, including the Hubble Space Telescope, has peered deep into the farthest reaches of the universe and discovered supermassive black holes residing in the centers of galaxies. Giant telescopes on Earth's highest mountaintops have spied planet-like objects larger than Pluto lurking at the very edges of our solar system and have carefully measured the expansion rate of our universe. Meteorites with bacteria-like fossil structures have spurred repeated missions to Mars with the ultimate goal of sending humans to the red planet. Astronomers have recently discovered hundreds more planets beyond our solar system. Such discoveries give us a reason for capturing what we now understand about the cosmos in these volumes, even as we prepare to peer deeper into the universe's secrets.

As a discipline, astronomy covers a range of topics, stretching from the central core of our own planet outward past the Sun and nearby stars to the most distant galaxies of our universe. As such, this set of volumes systematically covers all the major structures and unifying themes of our evolving universe. Each volume consists of a narrative discussion highlighting the most important ideas about major celestial objects and how astronomers have come to understand their nature and evolution. In addition to describing astronomers' most current investigations, many volumes include perspectives on the historical and premodern understandings that have motivated us to pursue deeper knowledge.

The ideas presented in these assembled volumes have been meticulously researched and carefully written by experts to provide readers with the most scientifically accurate information that is currently available. There are some astronomical phenomena that we just do not understand very well, and the authors have tried to distinguish between which theories have wide consensus and which are still as yet unconfirmed. Because astronomy is a rapidly advancing science, it is almost certain that some of the concepts presented in these pages will become obsolete as advances in technology yield previously unknown information. Astronomers share and value a worldview in which our knowledge is subject to change as the scientific enterprise makes new and better observations of our universe. Our understanding of the cosmos evolves over time, just as the universe evolves, and what we learn tomorrow depends on the insightful efforts of dedicated scientists from yesterday and today. We hope that these volumes reflect the deep respect we have for the scholars who have worked, are working, and will work diligently in the public service to uncover the secrets of the universe.

Lauren V. Jones, Ph.D.
Timothy F. Slater, Ph.D.
University of Wyoming
Series Editors

Preface

A great deal of attention has of late been directed to the smaller members of the solar system. In the past decade, we have seen spacecraft orbit and land on asteroids, and smash into comets. We have taken our first steps to better understand and characterize the threat to our civilization posed by collisions with near-Earth objects. For the first time in a generation, the sky was graced with extraordinarily bright comets. And scientists have begun to discover and catalog new planet-size objects in the far reaches of the outer solar system, leading to controversy about what exactly makes something a planet.

These new findings have been met with interest and enthusiasm by the general public. The prevention of and possible consequences of asteroid impact have been the central plot lines of popular movies, and the controversy over the status of Pluto has been immortalized in Web site petitions and T-shirts.

This increase in attention comes hand in hand with an increase in our knowledge about the small bodies of the solar system. The combination of new telescopes, more capable computer simulations, sophisticated laboratory techniques, and spacecraft data has made this a golden age for asteroid and comet studies. We have long known that these objects have much to tell us about the solar system, its history, and its evolution. We are only now able to understand large swaths of the story being told, and there is much more yet to be understood.

A volume in the Greenwood Guides to the Universe Series, *Asteroids, Comets, and Dwarf Planets* is intended as an introduction to the current state of knowledge; the volume is aimed at a nontechnical audience of students who do not plan to specialize in astronomy, geology, or physics, and at public library patrons who are interested in the field and seeking current information. The volume covers a wide range of topics of interest for these readers, if only at a level to whet a reader's appetite rather than provide in-depth knowledge. In contrast to a separate treatment of each population, the focus of this book is on processes and features and a comparison and contrast of how comets, asteroids, and dwarf planets are affected by those processes and to what extent they share those features. Also explained are the techniques and logic that have led planetary scientists to the conclusions that they have reached.

Containing 13 chapters and an Introduction to the topic, *Asteroids, Comets, and Dwarf Planets* begins by looking at the factors that set those objects off from the major planets. After a tour of the history of their investigation, we look at where the small bodies are found in the solar system, and what we have learned from the samples in our laboratories about their nature and formation. Following this come several chapters that look at the composition and structure of asteroids, comets, and dwarf planets, from their interiors to their surfaces, the tenuous atmospheres suspected on the largest objects, and the satellites that accompany some objects. Finally, we turn to the hazard posed by some of the small bodies and the different types of space missions that have been sent to them and what we have learned, before considering the interrelations between small bodies and other populations both inside and outside the solar system.

The chapters contain sidebars that offer more in-depth information on a broad array of important and engaging topics. The sidebars are included to give readers a deeper understanding of various issues and questions without interrupting the main flow of the chapter narrative. The chapters also contain numerous photographs and other illustrations that serve to augment the discussion in the text. Terms highlighted in boldface in the text alert the reader to the term's inclusion in a Glossary, which offers a brief, useful description of the term as currently used in the field. Each chapter concludes with a brief list of further print and electronic information sources for that topic; a bibliography at the end of the book lists more general information sources. The volume concludes with a detailed subject index.

Acknowledgments

I would like to acknowledge the help and support of a number of people without whom writing this book would have been more difficult and whose input was a big help.

Initial thanks go to the team at Greenwood Press (now ABC-CLIO)—Lauren Jones, Kevin Downing, and John Wagner. I greatly appreciate their patience, comments, and encouragement. I would also like to thank Tim Spahr, without whom I would not have had the opportunity to participate in this series.

My colleague Ralph Lorenz was a great resource in many stages of this process. His extensive writing experience greatly eased my burden by assisting me in identifying possible problems and their solutions well before I would have spotted them. My colleagues Neil Dello Russo and Nancy Chabot also contributed by lending research materials and answering my questions about the many specialty areas included in the book with which I have less personal experience.

Thanks also to my employers at the Johns Hopkins University Applied Physics Laboratory, particularly Andy Cheng, Ben Bussey, and Louise Prockter, for their support and encouragement.

Introduction

Current scientific research about asteroids, comets, and dwarf planets takes many forms. While one might imagine astronomers spending lonely nights at observatories on remote mountaintops, many of the latest advances have been made by someone using a laptop at their desk in a major city. It is the constant interplay between theory, modeling, observations, and experiments that makes planetary science in general, and small bodies studies in particular, so dynamic. It is important to note that despite the division of research into the following separate areas, many projects take advantage of more than one type of research and many scientists participate, for instance, in both observing and modeling. The bulk of *Asteroids, Comets, and Dwarf Planets* will focus on the current state of our knowledge of the small bodies, so we take a moment here to consider the general ways this scientific research is performed.

The first studies of the small bodies were done by observers, using their eyes alone, whose names are lost to history. In the intervening centuries, increasingly sophisticated instruments have been used in the discovery and study of comets, asteroids, and dwarf planets, first from the Earth, and eventually from space and even asteroid surfaces. At present, useful scientific observations are being made from several spacecraft and telescopes around the world, ranging from the largest that have been built to small homemade versions operated by hobbyists.

SPACECRAFT

The most visible and sophisticated means of learning about comets, asteroids, and dwarf planets is via space missions. Nearly a dozen small bodies have been visited by spacecraft, starting with a fleet of missions visiting Comet Halley in the 1980s. More missions are en route, including studies of the dwarf planets Ceres and Pluto in the coming decade.

These missions have provided surface imagery in great detail, which has been a boon to geologists and geophysicists seeking to understand the histories and processes on small bodies. They have also returned samples of cometary dust, and missions have been proposed to bring back surface samples of asteroids and comets. The quality of spacecraft data has revolutionized our knowledge of the small bodies.

OBSERVATION

Unfortunately, space missions are complex, expensive undertakings. Because of this expense and difficulty, only a handful of comets and asteroids have been visited by spacecraft, and the vast majority will never be explored in that way. Therefore, it is still the case that telescopes are used to collect most of the information we have for comets, asteroids, and dwarf planets. Well over 100,000 objects are known and catalogued, with several hundred thousand more having less well-known orbits. Most of these objects were discovered by Earth-based telescopes with diameters of 1–1.5 m. For comparison, the Hubble Space Telescope is 2.4 m in diameter, and the largest telescopes on Earth have sizes up to 10 m. Since the late 1800s, discoveries have been made using photographic techniques rather than by using an eyepiece, originally using film and moving to digital cameras in the 1980s.

Studies of the physical properties of small bodies can require larger telescopes, although there are many astronomy hobbyists who help advance science by making simple, but much-needed, observations on personally owned telescopes in their spare time. However, more difficult observations of fainter, smaller, and more distant objects use the largest facilities available to astronomers. Two of the more prominent facilities include the telescopes at Keck Observatory in Hawaii and the radar facilities at Arecibo Observatory in Puerto Rico.

EXPERIMENTS

While telescopic data is a critical component of the study of asteroids, comets, and dwarf planets, scientists have also been able to perform laboratory analyses and experiments to gain further insight. Meteorites, pieces of asteroids that fall to earth, are often the subjects of these analyses, as the most modern and sophisticated equipment available to geochemists determines their elemental compositions and identifies particular textures and minerals that give clues to the history of the meteorite and the entire solar system. Experiments using sophisticated furnaces and chambers simulate the temperatures, pressures, and compositions present when the meteorites formed. Experiments are also performed to study the effects of the solar wind and micrometeorite bombardment of asteroid and transneptunian objects on their apparent compositions. Some scientists perform experiments of a more energetic nature. These researchers seek to learn about impact processes in the solar system, and use guns designed to fire pellets at speeds of over 1 km/s (or over 3,600 km/hr). The lessons learned about the breakup of small objects are applied to the cratered and disrupted population of small bodies. In a sense, the *Deep Impact* mission was "just" a large-scale collisional experiment!

MODELING

Even small bodies are too large to allow experiments on the proper scale. For instance, the largest human-made craters on Earth have been made during nuclear tests and are still exceedingly small by astronomical standards. The effects of large impacts must be studied through mathematical extrapolation based on small craters and a knowledge of physics. The same sort of logic applies to studies of long-term processes that can take thousands to millions (or billions!) of years to complete.

The increased capabilities and availability of computing power has helped scientific research immensely. The small bodies community has turned this computing power to simulations of the gravitational interactions of large populations, leading to new theories about solar system formation, the migration of asteroids and comets through the solar system, the threat posed by the impact of near-Earth objects, and the formation and evolution of binary asteroids, to name a few applications.

THEORY

Finally, there are those whose research requires only a word processor or a whiteboard. The theoretical underpinnings of small bodies research are important areas and have shown particular advancement in recent years. The recognition that nongravitational forces like the Yarkovsky and YORP forces are important has flown from new theoretical appreciation. Consideration of early solar system history has led to new expectations for the properties and history of asteroids, comets, and dwarf planets. These expectations feed back into new proposed observations, experiments, and modeling efforts.

The following pages compile the results of decades of research, using all the techniques described herein. Science is always changing, in particular a field like small bodies studies, where a new meteorite fall or a bright comet or data returned from a spacecraft can be drastically different from previous experience. The aim of this volume is to provide a general guide to our current understanding, upon which new findings can be added and understood.

1

A Matter of Definition

According to the International Astronomical Union (IAU), the solar system contains one star, eight planets, at least five dwarf planets, and uncounted numbers of "small solar system bodies." According to others, however, there are nine planets in the solar system. Still others count at least 10 planets, if not more. In this chapter, we will attempt to define a "planet," or at least show how one might do so. We will also follow the evolution of this debate, and consider the pros and cons for each camp. We will then turn to the small bodies to lay a foundation for further discussions throughout this book.

How might we go about defining a planet? An admittedly circular way might be to use the objects we already call planets and try and figure out what they have in common. We might use characteristics like their orbits, sizes, and shapes.

For the ancient Babylonians and Greeks, little information was available, but one thing was clear—of the objects in the sky, most remained in place with respect to one another, while a small number moved. These moving objects, which we now call Mercury, Venus, Mars, Jupiter, and Saturn, as well as the Sun and Moon, were all considered "wandering stars," or *planetes asteres* in Greek. Comets were known to the ancients, but considered either supernatural or related to weather in some unknown way. As the heliocentric model gained acceptance and the Sun replaced the Earth at the center of the solar system, the two bodies swapped classifications, with the Earth being regarded as a planet, while the Sun was not. The word *planet* was further refined through the seventeenth and eighteenth centuries as objects were found orbiting Jupiter and Saturn; these objects were first called planets, but the term *satellite* eventually came into use for anything orbiting something other than the Sun, including our Moon. The discovery of Uranus in 1781 added a new planet to the scientific consciousness for the first time.

As described in more detail in Chapter 2, the first four of what we now call asteroids were discovered in the early 1800s. Each discovery was treated as a new full-fledged planet, and individual symbols were assigned to each body, just as had been done for the other planets. Observations quickly determined that this quartet was not like the rest of the planets, however. The first curiosity was the similarity of their orbits, which were much closer to each other than any other sets of planets. Furthermore, while most planets showed disks when viewed through telescopes, these four objects remained stubbornly starlike in appearance regardless of magnification. This fact led William Herschel to dub them "asteroids," meaning starlike.

By the 1850s, with over a dozen asteroids discovered, the general consensus was that the asteroids were fundamentally different from other planets, and they were considered their own class: **minor planets**. The major planets, by comparison, were large and orbited by themselves, along with any satellites.

Thus when Pluto was discovered in 1930, there was no question in anyone's mind that it was a planet. It was believed to have a significant mass and did not share its orbit with any other objects (even though it crossed Neptune's orbit, which attracted some notice). However, as we learned more about the planets and as estimates of Pluto's mass got smaller (and predictions for objects in the transneptunian region became stronger), astronomers began to question whether Pluto was really a planet.

A RULE-BASED DEFINITION?

When learning biology, students are presented with a few simple rules to determine whether something is a mammal: Does it give birth to live young rather than lay eggs? Does it feed its young milk? Does it have hair? Similarly, one could look at those objects that are undeniably planets and try and come up with a similar set of questions. However, this is not as straightforward as it may seem, as we can see from the following questions:

- *Does it have an atmosphere?* The giant planets, Venus, Earth, and Mars, all have atmospheres. It is arguable whether Mercury's thin sodium and potassium "atmosphere" should count in this way, though if it is considered to have one, so do a great many other objects including the Moon (as is further discussed in Chapter 10). There is no question that Saturn's satellite Titan has an atmosphere, however, although it is not generally considered to be a planet.

- *Does it orbit the Sun rather than another planet?* Many objects not normally considered to be planets orbit the Sun, such as asteroids and comets. This suggested rule would allow us to rule out Titan as a planet, but would still need to be combined with other rules.
- *Does it have any satellites of its own?* Again, there are objects known to have satellites that are not considered planets, for instance, the asteroid Ida (among others, as discussed in Chapter 6). Conversely, neither Mercury nor Venus have any satellites, and there is general consensus that they are planets.

Obviously, to try and define a planet in this manner will require several different questions to be answered, and agreement on which are the important questions and which are less useful. Perhaps there is another way to define a planet.

OBSERVATIONS AND DEFINITIONS: DOES SIZE MATTER?

Interestingly, the question, "What is a star?" does have a simple answer—A star is an object generating energy through nuclear fusion. Astronomers have been able to make observations of bodies and measure their temperatures, which, in turn, can easily let us calculate whether an object is a star.

Ideally, there would be a similar measurement or set of measurements that can be made to determine whether an object is a planet or not. For a long time, the mass of an object was informally treated as the discriminator between objects that are planets and those that are not. Above a certain size, about 13 times more massive than Jupiter, objects will undergo fusion and become stars. Planets, then, are 13 Jupiter masses or smaller. But what about the small end?

When looking at the sizes and masses of solar system objects, there is one place in the distribution where a break between planets and non-planets might naturally be placed since it is free of objects and it is unlikely that any objects will be found in the gap in the future—Uranus is 14 times more massive than the Earth and 4 times larger in radius. Some have argued that the classification that makes the most sense is for the solar system to have four planets: Jupiter, Saturn, Uranus, and Neptune, and everything else, including the Earth, is unworthy of the name! Few have ever taken this suggestion seriously, although that may be because as earthlings we are biased against it. Figure 1.1 shows the relative sizes of the planets (including Pluto); where might we draw a line?

The size of Pluto was long informally treated as the low-mass end for planets (again, with the usual restriction that larger-mass satellites like Ganymede and Titan were ineligible for planethood). The first reliable measurement of Pluto's mass came in the late 1970s, after the discovery of its satellite Charon. Pluto's mass was found to be by far the smallest of any planet, including Mercury. Through the 1990s and early 2000s, a group of objects orbiting beyond Neptune and called **transneptunian objects (TNOs)** were

Figure 1.1. The nine bodies generally recognized as planets in the late twentieth century have a wide range of sizes, shown here to scale (although their distances from one another and the Sun are not to scale). Pluto, at the far right, is much smaller than any of the others, and is much more similar in size to large transneptunian objects discovered in the past 10 years. The difference in size between the jovian planets (Jupiter, Saturn, Uranus, and Neptune) and the terrestrial planets (Mercury, Venus, Earth, and Mars) is as striking as the difference between Pluto and Mercury, the smallest terrestrial planet. NASA/Lunar and Planetary Laboratory.

found with sizes (and estimated masses) ever closer to Pluto's, until finally the discovery of Eris showed that objects larger than Pluto were still to be found. Was Eris the tenth planet? Or did this show that Pluto was simply the first TNO to be discovered and not worthy of a planetary classification? The answer depends in a lot of ways on what you think a planet *is*, which is the question we seek to answer in the first place.

What's in a Name?

The naming of various planets and dwarf planets has caused a surprising amount of controversy and discord over the years. In the seventeenth century, Galileo proposed to honor his patrons by naming what are now called the Galilean satellites of Jupiter "the Medicean planets" after the Medici family. Similarly, in the eighteenth century, William Herschel called his discovery "*Georgium Sidus*" (The Georgian Star) after King George III of England. The French, understandably reluctant to name a new planet after the king of England, preferred the name "Herschel," with the German preference of "Uranus" eventually winning out. Ceres was originally called "Ceres Ferdinandea," in partial homage to King Ferdinand III of Sicily. Here too, the reference to an earthly ruler was omitted. Finally, the discovery of Pluto led to an outpouring of suggested names, including Minerva, Cronus, and the somewhat unlikely Constance (proposed by Constance Lowell, widow of Percival Lowell, astronomer and founder of the Lowell Observatory, where Pluto was discovered).

Naming controversies have continued to this day among the asteroids. Particular uproar faced the naming of 2309 Mr. Spock, named not after the Star Trek character, but after the discoverer's cat. This led to a ban on asteroids named after animal pets. The naming of Eris (goddess of discord) and its satellite Dysnomia (lawlessness) was a nod toward the classification controversy they helped bring to a head.

WHY DEFINE?

The potentially circular nature of defining planethood is considered an unsolvable problem by many astronomers. They don't feel that a definition is necessary, arguing that any distinction between planets and non-planets is arbitrary and not important scientifically. The word *planet* already has a well-established meaning in many languages, and while it may not be a precise analytical meaning, it is not, astronomers argue, their place to redefine this word.

An example of this viewpoint is to think of the definition of a book. If asked to describe a book, many people will come up with similar descriptions. However, the details may disagree. Some will argue that a book must have content of some kind, so blank books and phone books are not books. Others would require books to be physical objects, so an electronic book would not count. Nevertheless, most people could identify most things they encountered as either a book or not a book. Similarly, some would argue, the details of what is a planet don't matter, because scientists basically agree whether a specific object is a planet even if they don't all agree exactly on the details of the classification. They see further parallels in the definitions of life or art (or obscenity—as a U.S. Supreme Court justice once noted, he wouldn't define obscenity, but he "knew it when he saw it"), which at their extremes have no consensus.

On the other hand, supporters of some kind of definition note that many words in common use also have specific scientific definitions, such as *sand*. They argue that to reduce confusion in scientific work, a definition is critical.

THE SHAPES OF THINGS

For many astronomers, the importance of a definition is that it should quickly allow a determination to be made for an object, with no ambiguity (at least in principle). This suggests a reliance on observable properties such as brightness, mass, size, or orbit.

The shape of an object has been used as a basis for classification. The largest objects in the solar system are basically spherical objects. Why is this so? We have all heard of "sea level." The oceans of the Earth have surfaces at the same relative height. Water in rivers and streams on the Earth's surface flows downhill until it reaches the ocean and sea level. If somehow the surface of the Atlantic Ocean were higher than the Pacific and Indian Oceans, water would flow "downhill" out of the Atlantic and into the other oceans until they were once again at the same level. If water were added to the oceans, via melting ice or other means, the water would distribute itself around the world until a new level was reached, higher in comparison to the land but still equal across all the oceans.

Imagine a planet with no dry land at all, only water. Everything at its surface would be at the same level—sea level. The distance from any point on its surface to the center of the planet would be the same. In other words, the planet would be spherical. A planet does not have to be covered in water for this to be true, however. A fluid planet with a thick atmosphere and no real solid surface, like Jupiter or Saturn, will also tend to be spherical for similar reasons.

Even solid bodies like the Earth will become spherical. The large amount of mass in the Earth can act like a fluid over a course of years. As a result, the difference in elevation from the highest point on the Earth's surface (Mt. Everest) to the lowest point under the ocean (in the Marianas Trench) is less than 0.2 percent of the mean radius of the Earth. By these standards, the Earth is very smooth indeed. Mars, with higher peaks and deeper valleys, is still smooth within 1 percent of its radius.

Planets are not *exactly* round, however. Precise observations of Jupiter, or even the Earth, show that their exact shapes are **oblate**, with the distance from north pole to south pole slightly less than the equatorial diameter. This is because planets rotate. For instance, all objects on the Earth have an angular speed of 360 degrees per 24 hours (in other words, they complete one rotation in a day). The linear speed can vary dramatically, however; objects on the equator travel on a circle with a circumference of roughly 40,000 km every day with respect to the center of the Earth, nearly 1,700 km/h. Objects at the north or south poles, however, don't move at all with respect to the center of the Earth. Objects at various points between the Equator and poles move at intermediate speeds depending on their latitudes. The extra speed felt by objects on a planetary surface due to rotation causes an extra force, which makes material flow toward the equator, causing planets to be not-quite spherical. Objects that act in this way are said to be in **hydrostatic equilibrium**. Large masses like the planets, and smaller fluid masses, are found to be in hydrostatic equilibrium. Smaller solid masses, however, cannot overcome their own strength and can retain irregular shapes (like the asteroids Ida or Eros, among many others).

Whether an object is massive enough to be in hydrostatic equilibrium is seen by many astronomers as a key criterion for planethood. For massive enough objects—like Jupiter, Neptune, or Venus—hydrostatic equilibrium is guaranteed. At the small end, observations might be necessary to determine whether an object is massive enough (or is the right shape, if mass measurements are not available) to be in hydrostatic equilibrium.

A proposal was made in 2006 to the International Astronomical Union, which would have defined a planet as an object that had the following characteristics:

1. Sufficient mass to be in hydrostatic equilibrium
2. Orbited a star
3. Is neither a star itself nor a planetary satellite

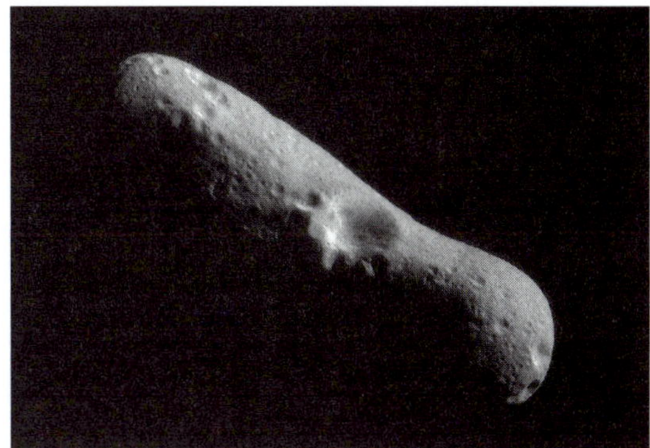

Figure 1.2. Two "non-planets" appear in this figure. At left is Saturn's satellite Titan, which is larger than the planet Mercury and has a thick atmosphere. It is large enough to be in hydrostatic equilibrium, one of the criteria for planethood in the recent IAU definition, but because it is in orbit around Saturn, it is not eligible. At right is the asteroid Eros, visited by the *NEAR Shoemaker* spacecraft. It is much smaller than Titan, too small for its gravity to have pulled it into a spherical shape. It is also much too small for its gravity to affect other objects in nearby orbits. It therefore fails both tests for planethood. However, classification as a non-planet does not diminish the scientific value and interest in these or the other large satellites and small bodies of the solar system. Left, NASA/JPL/Space Science Institute; right, NASA/JPL/JHUAPL.

In addition, a category of "double planet" was proposed, if two components of a system were each in hydrostatic equilibrium, had similar masses, and a particular orbital configuration. Pluto and Charon would have qualified as a double planet in this definition. In addition, Ceres and Eris would have qualified as planets, giving the solar system 12 known planets. A dozen or more additional objects, almost all beyond Neptune, might also have been classified as planets, pending the collection of additional data.

This proposed definition did not satisfy everyone. Major and minor problems were pointed out, with some people uncomfortable with the idea that the solar system might have dozens of planets. Others argued that studies showed Pluto to be a typical TNO and convenience and history were no reason to confer special status, regardless of its shape.

A DYNAMICS-BASED CLASSIFICATION

What would be a reason to confer special status? One proposed classification looks to the formation of the solar system, discussed in further detail in Chapter 5. The eight undisputed planets from Mercury to Neptune all were very effective in either collecting (or **accreting**) mass, capturing it into satellite orbits, or ejecting it to other parts of the solar system (or out of the solar system entirely). By the end of solar system formation, the vast majority of material near the Earth's orbit was part of the Earth itself. The same is true for the other planets, from Mercury to Neptune. This process, informally dubbed "clearing of neighborhoods," is considered by some a critical test of planethood.

A quantitative measure of the ability of an object to clear its neighborhood is provided by a parameter called Λ (the Greek capital letter lambda). This is equal to an object's mass (M) squared times a constant (k) divided by its orbital period (P): $\Lambda = kM^2/P$. We can compare different objects by measuring masses in Earth masses and the period in years, setting $k = 1$. Then $\Lambda = 1$ for the Earth. Jupiter, over 300 times the mass of the Earth and with a period of nearly 12 years, has a value of Λ roughly equal to $(300)^2/12$, or 7,500. Neptune's mass is 17 times that of the Earth, its orbital period is just short of 165 years, and its Λ is roughly 1.8. For Mars, $\Lambda = 0.006$, apparently quite small but still above the threshold expected for neighborhood clearing.

Pluto, which orbits relatively close to other objects of similar mass, fails this test. Its mass is only two thousandths that of the Earth, and its period is nearly 250 years. As a result, Λ for Earth is 50 million times larger than for Pluto, not nearly enough for Pluto to have any real dynamical effect on nearby objects. Dynamicists, who wish to study the orbits of asteroids and comets over millions of years or longer, need to include the gravitational effects of the major planets to have accurate results, but find that they can ignore Pluto in their studies without any loss of accuracy.

Using the criterion that an object must have "cleared its neighborhood," as well as the previous requirements for hydrostatic equilibrium and not orbiting another planet, we are left with eight planets: Mercury, Venus, Earth, Mars, Jupiter, Saturn, Uranus, and Neptune.

This dynamics-influenced definition has been adopted by the International Astronomical Union. It also created a category called **dwarf planets**, which are objects that are large enough to be in hydrostatic equilibrium, but are not capable of clearing their neighborhoods. This category originally included Pluto, Ceres, and Eris, although two additional objects have been officially included into this category since the original announcement, and more objects could be classified as dwarf planets pending additional data. An additional category was added to the IAU classification scheme in 2008, with dwarf planets orbiting beyond Neptune being further classed as **plutoids**. Thus Eris, for instance, is both a dwarf planet and a plutoid, while Ceres is only a dwarf planet. Everything else in the solar system that orbits the Sun rather than a planet is then classified as a **Small Solar System Body (SSSB)**.

THE UNSETTLED SETTLEMENT

While there is now an official IAU definition of a planet, it is not certain as of this writing that the last word has been spoken. The most common concern with the official definition is that it is vaguely worded, and based on models and simulations in a critical way, and those models could change with time. A definition based only on observations, like the ones based only on shape, would not suffer from that drawback. In addition, it has been noted that using the IAU definition and the models for neighborhood clearing, an object the size of Mars found at the distance of Eris would not qualify as a planet, and indeed a hypothetical Earth-sized object at a large enough distance from the Sun would not be considered a planet using that criterion either.

An additional complaint is that by the technical wording of the IAU definition, technically speaking there are no planets outside the solar system. It would be possible to change some of the language to allow planets around other stars, but without knowing the details of those systems, it would be difficult or impossible to determine whether any of those objects have cleared their neighborhoods, and difficult or impossible to apply the IAU definition. It is also hard to see how any "free floating planets," planet-size objects that may have escaped the solar systems in which they formed early in their history, could be accommodated by a definition that includes dynamics. All told, it appears likely that the IAU definition of a planet will be refined in the coming years, and it is possible that researchers will continue to use definitions that do not match any official definition at all.

AMONG THE SMALL BODIES

While the definitions of groups of small bodies are not as contentious as those for planets, they are in some ways no less murky. Asteroids and comets have been distinguished from one another for centuries by a simple observational question: Does the object appear starlike or "fuzzy" through a telescope? As noted before, asteroids were named for their starlike appearance, and are **point sources** (that is, they are so small they appear as points of light regardless of magnification) in all but a very few telescopes, including the Hubble Space Telescope. Comets were named for their "fuzzy" comae and tails, the word *comet* meaning "hairy star" in ancient Greek. While the ancient Greeks only had their eyes for observing, this distinction between asteroids and comets has been maintained to date, with any object showing evidence of a coma or tail being classified as a comet. However, some objects have been found to sometimes appear starlike and sometimes have a coma. If an object appears cometary when discovered, it will maintain this classification. If, however, an asteroidal object is later seen to have a cometary appearance, it can be given a cometary classification as well as an asteroidal one. To date there are three such objects.

By the late-twentieth century, it was recognized that comets were largely icy bodies that originated in the outer solar system, and asteroids were largely rocky bodies that originated in the inner solar system (see Chapter 5). This has led to a separate, informal distinction between comets and asteroids. For some researchers, *asteroid* has become shorthand for a rocky object, and *comet* for an icy one. Alternately, asteroids represent bodies that originated in the inner solar system, and comets in the outer solar system.

Unfortunately, these two distinct ways of separating comets and asteroids can lead to occasional confusion. One will sometimes hear of "dead comets," which show no evidence of a coma or tail (and thus would be asteroids by the observational definition), but are believed to have originated in the outer solar system (and would thus be comets by the compositional/dynamical definition). In the chapters that follow, we will use the compositional/dynamical definition unless otherwise noted—asteroids are generally rocky and metallic objects believed to originate in the inner solar system; comets are generally icy bodies believed to originate in the outer solar system. In some cases, the line is blurry, particularly for some populations of bodies for which we currently know little. We will address the relationship between asteroids and comets, and the objects that straddle the lines between them, in Chapter 13.

Finally, another term we will often use in later chapters is transneptunian object. As mentioned previously and suggested by their name, TNOs are found beyond the orbit of Neptune. These are thought to be similar in composition to comets in general, but are named to indicate that their current orbits are far from the Sun.

SUMMARY

The classification of the bodies of the solar system is a surprisingly subjective process. There are several different underpinnings for defining planets and non-planets, each with strengths and drawbacks, and no objectively "correct" answer. The diversity of the small body population is reflected in the diversity of how we classify them—comets and asteroids are distinguished from one another by composition and dynamical history, as well as by centuries-old observational descriptions. The new category of dwarf planet is based on shape and theoretical models of gravitational influence. TNOs are defined based on their orbits.

WEB SITES

Scientific American article about the "Planet Debate": http://www.sciam.com/article.cfm?chanID=sa006&articleID=93385350-E7F2-99DF-3FD6272BB4959038&pageNumber=2&catID=2.

A history of the status of Pluto: http://www.cfa.harvard.edu/icq/ICQPluto.html.
Video of IAU debate: http://www.astronomy2006.com/media-stream-archive.php.
Commentary on the IAU debate from a dissenting astronomer: http://www.lowell.edu/users/buie/pluto/iauresponse.html.
NASA page about the definition of planets and dwarf planets: http://solarsystem.nasa.gov/planets/profile.cfm?Object=Dwarf&Display=OverviewLong.

2

Historical Background

The small bodies of the solar system have been of interest to astronomers for centuries. While we continue to learn more about them with each passing year, in this chapter we will focus on the earliest stages of their study. We will consider how asteroids and comets were interpreted and how those interpretations were shaped by then-current theories, and how they in turn shaped how those theories evolved.

OMENS AMONG THE WANDERERS

Astronomy is a modern science with ancient roots. Our ancestors were familiar with the night sky. Living without electricity in relatively small groups, and critically dependent upon understanding the seasons in order to plant and reap their crops at the right times, they gained an intimate knowledge of the cycles overhead. They could see thousands of stars, whose rising and setting times slowly changed through the year, but whose positions relative to one another were unchanging.

They also knew of lights in the sky that moved relative to the stars. Among them were the two brightest objects in the sky—the Sun and the Moon. In addition, there were five bright objects that appeared like stars to their naked eyes but did not remain fixed. These "wandering stars," called "planets" in Greek, are known to this day by the names of Roman gods: Mercury, Venus, Mars, Jupiter, and Saturn. The stars and these seven wanderers acted in predictable ways, and the ancients of many cultures tracked them day after day, year after year, and century after century.

Occasionally, however, this predictability was unexpectedly disturbed. The stargazers of the time, who believed studying the movement of the stars and planets could lead to an ability to predict the future (a nonscientific belief that has evolved into modern astrology), were often rattled by these disturbances, and they connected floods, wars, epidemics, famines and the likes to these "bad stars" or "disasters" in Latin.

The most common of these bad omens were the objects we know as comets. Appearing without warning and looking like brooms, or fiery swords, comets were often noted in conjunction with the deaths of kings, including Genghis Khan and Julius Caesar. A comet seen in 66 AD was believed to be an omen of Jerusalem's destruction, which occurred only a few years later. Perhaps the most famous association of a comet with the fates of kingdoms was the appearance of Comet Halley in 1066. Harold, King of England, saw the comet as a bad omen, while it raised the morale of William of Normandy and his successful invading troops. The events of the Norman invasion of England were recorded in the Bayeux Tapestry, which includes depictions of the comet.

COMETS AND CRYSTALLINE SPHERES

As the modern era of Western science began, observations of comets began to be used to learn quantitative facts about our universe rather than seen as portents of good or ill fortune. After the sixteenth century, when Copernicus suggested that the Earth orbited the Sun, rather than vice versa, other astronomers continued the work of trying to figure out how the stars and planets actually moved. Without a conception of gravity, it was thought since the time of the ancient Greeks that the stars and planets all were physically attached to transparent spheres (often called "crystalline spheres"), with one sphere per planet and an additional one for the stars.

The existence and behavior of comets directly challenged the idea of crystalline spheres—not only would each comet require its own sphere, but comets seemed to move over a variety of distances, making it seem that they would smash into the other planetary spheres during their visits. For this reason, many astronomers concluded that comets were not astronomical objects, but were simply an unusual type of long-lived weather, generated in the Earth's atmosphere and just appearing to be among the planets. This view was popularized by Aristotle around 350 BC (who also first named these objects *kometes*, meaning hairy star in ancient Greek, after their appearance).

In the late sixteenth century, Tycho Brahe, the greatest astronomer of his era, performed painstaking observations of the sky. The precision of his measurements, made in the last decades before telescopes were invented, allowed him to measure the **parallax**, or apparent angular shift, of celestial bodies. An easy illustration of parallax is as follows: Close one eye and hold your thumb so it is blocking an object across the room from you. Now switch the eyes that are open and shut, and notice that your thumb is no longer blocking the object. The apparent motion of your thumb is greater if it is held closer to your face.

This same principle can be used for objects in the sky. The Moon, for instance, will be seen in a slightly different position relative to the stars when viewed on opposite sides of the Earth. In Tycho's time, collaboration with Chinese or Islamic astronomers was not possible, but because of the Earth's rotation, Tycho could make all of the necessary observations himself and remove the effect of the Moon's rotation to obtain the correct answer. A comet that appeared in 1577 provided another object for which Tycho could measure parallax (see Figure 2.1). However, he found no measurable parallax, implying a distance beyond the Moon and certainly outside of the atmosphere. These observations and conclusions showed that the crystalline spheres could not exist.

Tycho died before telescopes were used for astronomy, and the Comet of 1618 was the first to be observed with the new tool. Galileo Galilei,

Figure 2.1 This woodcut illustrates the Great Comet of 1577, shown by Tycho Brahe to be a solar system body rather than a weather phenomenon, as comets were thought to be by scholars of the time. This comet is now more than 10 times further from the Sun than Neptune is. Joerg P. Anders/Bildarchiv Preussischer Kulturbesitz/Art Resource, New York.

perhaps the most famous astronomer of all time, made observations. However, the first telescopic observations are usually credited to Johann Baptist Cysat, a Swiss Jesuit astronomer. Cysat was the first to distinguish the nucleus of a comet, and demonstrated that the comet's orbit was more parabolic than circular, indicating that its aphelion was exceedingly distant.

TELESCOPES AND COMET HUNTERS

By the end of the seventeenth century, great advances in astronomical theory by Johannes Kepler and Isaac Newton established the Universal Theory of Gravitation, and eliminated the need for any crystalline spheres. Newton's friend Edmund Halley noticed a pattern when looking at cometary appearances, and a similarity between comets that appeared in 1456, 1531, 1607, and 1682. Halley hypothesized that comets could be in orbits that made relatively frequent returns, and that all of those comets were in fact one and the same. He predicted this comet would return in 1758, a correct prediction that would lead to the comet taking his name: Halley's Comet, known as Comet Halley to astronomers.

Around this same time, telescopes began to be used in making new comet discoveries instead of merely making observations of known objects. Dedicated searches of the sky had been performed for centuries by Chinese astronomers, but the invention of the telescope now enabled Europeans to make discoveries much more frequently. The astronomical pastime of "comet hunting," popular among hobbyists to this day, had a number of immediate effects on both astronomical culture and astronomical paradigms.

In the first case, a number of astronomers found that some objects in the sky looked like comets but were fixed among the stars. One particularly active comet hunter, Charles Messier, compiled and published a list of objects that were prone to be confused with comets in the relatively small telescopes available in the 1770s. This list, now known as the Messier Catalog, was the first collection of what were simply called "nebulae" (or clouds), but are now known to include galaxies, globular clusters, and planetary and star-forming nebulae. Thus, comet hunting led to the first study of these deep-sky objects.

Comet hunting also cemented a realization that there were potentially many objects in the universe that were yet undiscovered. In 1781, William Herschel found what he thought was a comet, although further observations revealed an exceedingly distant orbit, no tail, and a disk. Rather than a comet, Herschel had discovered the planet Uranus—the first planet to be discovered in human history. There is evidence that Galileo observed the planet Neptune when it passed close to Jupiter in 1612, and he even noted that it moved, but since there was no expectation that any unfound planets existed (and indeed, Galileo still had lingering doubts about Tycho's conclusions about comets), Galileo did not realize what he had found. By Herschel's time, astronomers accepted the potential for yet-undiscovered objects. When the discovery of Uranus made Herschel a famous man, it was met with acclaim rather than skepticism.

THE CELESTIAL POLICE

As astronomers began to have a sense of the relative distances between the planets, they noticed a surprisingly large gap between Mars and Jupiter. By the mid-eighteenth century, it was recognized that the distances to the planets followed a pattern:

$$a = 0.4 + 0.3 \times 2^m$$

where a is the **semi-major axis**, or average distance from the Sun, and m is the set of nonnegative integers, plus negative infinity. Given this odd-looking set of rules, the following distances result:

M	Predicted A	Planet	Actual A
$-\infty$	0.4	Mercury	0.39
0	0.7	Venus	0.72
1	1.0	Earth	1.00
2	1.6	Mars	1.52
3	2.8	??	
4	5.2	Jupiter	5.20
5	10.0	Saturn	9.54

This pattern, called **Bode's law** or the **Titius-Bode relation** after the mathematicians who popularized it, matches planetary distances well, but some astronomers wondered about the "missing planet" that "should" be at 2.8 **astronomical unit (AU)**, where 1 AU is defined as the mean distance between the Earth and Sun. The discovery of Uranus furthered the suspicion that additional planets may be lurking unseen in the solar system. In addition, Uranus's distance from the Sun is pretty well predicted by Bode's law with $m = 6$. All of this evidence led a Hungarian astronomer, Baron Franz Xaver Von Zach, to propose a systematic search for the "missing" planet in 1800. He compiled a list of European astronomers who might join the effort, intending to assign a search area to each one and nicknaming the group the "Celestial Police."

As the invitations were making their way through war-torn Central Europe, however, a discovery was already being made. Giuseppe Piazzi, a

Sicilian astronomer, was mapping the sky to help comet hunters make more precise measurements. While taking positions of stars in the constellation Taurus on January 2, 1801, he noticed that one of the star positions had changed from the previous night. Further observations on the third and fourth nights confirmed his suspicions. The new object didn't show any evidence of a coma or a tail, and its motion was more typical of a planet's motion than a comet's. Nevertheless, Piazzi conservatively announced that he had found a comet and hoped further observations would prove his greater hopes correct. Unfortunately, an untimely illness confined Piazzi to bed, and by the time he recovered this object had moved too close to the Sun to be observable.

Piazzi's observations were not sufficient to allow an orbit to be calculated, given the types of orbital calculations available at the time. Worse still, that meant that there was little chance of finding it when it again became observable, except via luck. Because of the possibility that Piazzi's object was the missing planet, there was great interest in using the observations available to calculate a position, or at least to narrow down the search area to as small as possible. Carl Friedrich Gauss, a young mathematics prodigy from Hanover, turned his attention to the problem and developed new methods for calculating orbits, many of which are still used today. He calculated a predicted position for later in the year, and on the last day of 1801 Piazzi's object was recovered by Von Zach. The new positions allowed further improvements and greater certainty in the object's orbit, showing that it could indeed be considered a planet and that its orbit was where Bode's law predicted it should be. A delighted Piazzi named his object Ceres Ferdinandaea after the Roman-era patron deity of Sicily (Ceres) and the then-current king of Sicily, Ferdinand. In time, the object became known simply as Ceres. As 1802 opened, therefore, astronomers were content that they had found the "missing" fifth planet. They were in for a surprise, however.

In March 1802, Dr. Heinrich Wilhelm Olbers of Bremen intended to make observations of Ceres in hopes of improving the orbit of the newly discovered "planet." Instead, only a few degrees away from where Ceres was supposed to be, he found another object of roughly the same brightness. This new object proved to have a very similar orbit to Ceres, except for an orbital plane very different from the known planets. Olbers named this new planet Pallas, and astronomers struggled to make sense of the situation. Within five years, two additional objects were discovered. In recognition of Gauss's contribution to astronomy, he was allowed to name one of the planets, choosing the name Vesta. By this time, however, it was clear that the four new planets were quite different from the others. William Herschel's observations of Ceres and Pallas showed that they were many times smaller than Mercury or the Moon. Even at the highest magnifications with his largest telescope, they still had starlike appearances in contrast to the other planets, all of which showed measurable disks. For this reason, Herschel suggested that Ceres and Pallas (and later Juno and Vesta) really were a different class of objects, which he named "asteroids" for their starlike appearance.

PIECES OF PLANETS

It was Olbers who first suggested an origin for the asteroids, which had a profound effect on popular culture, if not an enduring acceptance in science. Noting the small sizes of the asteroids, he wondered if their origin could be explained as fragments of a normal-sized object that was disrupted or exploded. This led astronomers to wonder whether there were further asteroids to discover. Although there is ample evidence against this origin for the asteroids, its legacy remains in the creation story of Superman and the debris left after the Death Star destroyed Alderaan.

Ironically, the best support available in the early nineteenth century for the exploded-planet hypothesis was published in the decade before Ceres would be discovered, and further studies would show that asteroids could not have originated on a single disrupted body.

Stones and pieces of metal falling from the sky have been reported for centuries. As of the late 1700s, these were thought to be rocks flung from distant volcanoes (even as distant as the Moon!) or perhaps created in the Earth's atmosphere. More prevalent still was the conclusion that the reports themselves were mere folktales not to be taken seriously.

In 1794, however, Ernst Cladni published a book that reassessed these reports and proposed that they could best be explained as due to objects entering our atmosphere from outer space, often creating fireballs during entry. Cladni's case was bolstered by the newly emerging disciplines of geology and chemistry, and the lucky occurrence of several falls over the next decade. Chemical analyses of the fallen irons and stones showed that they both contained metallic nickel, which is not seen in terrestrial rocks. The discovery of Ceres, Pallas, Juno, and Vesta provided further evidence that small objects could be present in-between the planets, though it was over a century before asteroids and meteorites were conclusively associated with each other.

NEW TECHNIQUES AND DISCOVERIES

Just as Pallas was discovered by someone looking for Ceres, the fifth known asteroid, Astraea, was discovered in 1845 by someone looking for Vesta, the fourth. This discovery, and the discovery of three more asteroids in 1847, led astronomers to reconsider the status of objects in the asteroid belt. The 39-year gap between the discoveries of Vesta and Astraea would be the longest drought for asteroid hunters, and Figure 2.2 shows how the number of known asteroids grew quickly through the second half of the nineteenth century.

This increase in discoveries was accompanied by an appreciation of the potential usefulness of asteroid studies by some mathematicians. By 1870, Kirkwood had shown that the orbits of the 100 known asteroids spanned a relatively wide range of distances from the Sun. However, the range of solar distances covered by the asteroids was not uniform and avoided regions where Jupiter's pull was disproportionately strong because of **mean motion resonances**, where the orbital period of an asteroid was exactly one-half,

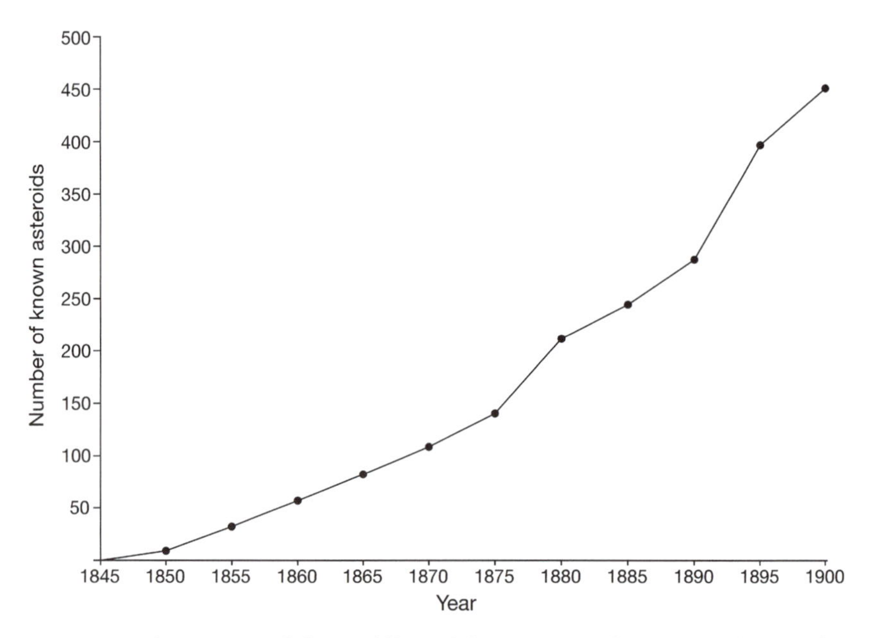

Figure 2.2 Over the course of the middle and late nineteenth century, the number of known asteroids mushroomed from four to more than 450. The pace of discovery was relatively slow at first, but the introduction of photography led to an explosion of findings, both via dedicated searches and accidental observations by astronomers studying stars and galaxies. Illustration by Jeff Dixon.

one-third, or two-fifths the period of Jupiter (among other similar fractions). Kirkwood argued that this was inconsistent with their origin in a single destroyed planet, but was what one might expect if the planets formed in a disk of gas and dust. This origin scenario is similar to the currently accepted theories for the formation of the solar system (see Chapter 5).

By 1870, astronomers were also starting to take advantage of the new technology of photography. Henry Draper and others first took pictures of the Moon and Sun, following with pictures of stars. In the 1880s, Draper took the first successful picture of a comet, and Max Wolf was the first to use photography to discover an asteroid. Wolf went on to discover over 200 asteroids beginning in the 1890s. The use of photography enabled a permanent record to be made of observations, with a much higher accuracy than was available from sketches made by observers looking through telescopes with their eyes. Photography was also combined with the new technique of **spectroscopy**, where the target's light is split into its constituent colors, which vary depending on composition, as described in greater detail in Chapter 7. Spectroscopy of comets showed that they have **organic material** (containing both carbon and hydrogen) on their surfaces.

"THE VERMIN OF THE SKIES"

In the closing decades of the nineteenth century, astronomers used their newly developed tools like spectroscopy and photography and their new large telescopes to study the universe outside of the solar system, which was available to them for the first time. As a result, interest in small bodies research lagged. Indeed, as the new field of astrophysics was maturing, asteroids in particular were often considered only as unwanted objects that littered their photos of nebulae and galaxies and were sometimes derided as "vermin of the skies." This attitude was compounded by the small sizes of the newly found asteroids. Furthermore, while the earliest estimates of the sizes of asteroids were typically overestimates, the new technique of **photometry** led to large underestimates of their diameters. Photometric observations measure the brightness of objects, which is a function of their distance as well as the fraction of light they reflect, or **albedo**. The albedos assumed for the large asteroids were poor estimates, and the sizes were wrong by a factor of two. By this time, the term *minor planet* began to be used for the asteroids and comets. Ceres, Pallas, and Vesta, with similar orbits to the hordes of new discoveries and deceptively small size estimates, were seen as much more similar to their smaller cousins than Mars or Mercury. By the second half of the 1800s, the largest asteroids had lost their status as planets.

However, some astronomers and mathematicians still found asteroids and comets interesting because of their orbits. In 1898, 433 Eros was discovered and was of immediate interest because of its proximity to Earth and status as the first **near-Earth object (NEO)**. It was realized that observations of Eros could take advantage of the same principle of parallax used by Tycho in the sixteenth century. The technology available in the early 1900s allowed Eros's parallax to be measured, providing a value for the distance between Earth and the asteroid, which was in turn used to calculate the distance between the Earth and the Sun. This was the most accurate measure of that distance, known as the AU, available for many years.

The discovery of Eros also inspired two advances that have come to be routinely applied to small bodies to this day. The first was the finding that Eros's brightness varied with time in a repeating pattern. It was quickly realized that this brightness was most likely due to Eros's rotation and that by observing the change in brightness, or **lightcurve**, Eros's shape and the direction of its rotation axis could be calculated. Lightcurve studies have been mainstays of asteroid and comet observations ever since.

The second advance was spurred by the desire to refine Eros's orbit. Vast collections of photographs of the sky were kept at observatories throughout the world. After Eros's orbit was first calculated, dynamicists realized that it was probably present on some of these old photographs, and located its position. These "precoveries" have been used to improve NEO orbits, particularly in cases where objects are potentially on a collision course with the Earth.

Within a few years of the discovery of the first NEO, a discovery was made on the other side of the asteroid belt. Among the hundreds of asteroids found photographically by Max Wolf was an object that shares an orbit with Jupiter. This object, 588 Achilles, orbits near one of the **Lagrangian points** of Jupiter. Objects in those orbits are protected from perturbations by Jupiter's gravity. Achilles was only the first of thousands of objects, collectively called **Trojan asteroids,** now known in one of these orbits. Mars and Neptune are also known to have objects at their Lagrange points.

The increased pace of discoveries to close the 1800s meant that studies could be done on the entire population of asteroids, looking for trends and patterns. This type of study was first done by Kirkwood, mentioned previously. In 1918, the Japanese astronomer Kiyutsugu Hirayama identified close orbital similarities between some groups of asteroids. He correctly deduced that these groups, which he called families, could have formed from the breakup of objects with the fragments remaining in similar orbits. These groups are now called **dynamical families**, or **Hirayama families** after their discoverer. Dynamical families are still under study today, as they provide unique opportunities to observe the effects of disruption and how the surfaces of asteroids change with time.

KEEPING TRACK OF TRACKS

With the increase of asteroid discoveries came an increase in the number of orbits to catalog. In order to determine which observations were discoveries and which were of already known objects, orbital calculations were required for each known asteroid. The Astronomical Calculation Institute in Berlin began this work in the 1870s, cataloging the known observations and adding new observations to try and recover objects that had already been lost (an ongoing project that stretched into the year 2000). The Berlin group was also the first to publish alerts soliciting observations so that newly found objects would not also join the ranks of the lost.

This work, interrupted by World War I, was officially sanctioned by the new International Astronomical Union (IAU), which was formed in 1919 by a coalition of different national groups of astronomers. After World War II, the IAU reassigned the work that had been done in Germany, with astronomers in the Soviet Union responsible for publication of up-to-date asteroid orbits, and the Minor Planet Center in Cincinnati responsible for the alerts. The Minor Planet Center moved to Cambridge, Massachusetts, in the 1970s, but still publishes "Minor Planet Circulars" via e-mail with the latest discoveries and calls for observations.

THE SEARCH FOR PLANET X

At roughly the same time as the discovery of Astraea, astronomers noted discrepancies in the position of Uranus. That planet had completed most of an orbit since its discovery. More detailed calculations of its orbit were made to try and understand these discrepancies, taking into account the effects of

Jupiter and Saturn. The French astronomer and mathematician Urbain Le Verrier concluded that an additional planet must exist and was perturbing Uranus's orbit. Le Verrier predicted a position for the new planet, and it was later found close to the predicted position. This planet was eventually named Neptune. Le Verrier's achievement was much celebrated; it was the first successful prediction of a planet based on theoretical principles, as opposed to the accidental discoveries of Uranus and the asteroids.

Some 60 years later, there appeared to still be small discrepancies in the positions of Uranus. Percival Lowell, founder of Lowell Observatory, used the same mathematical techniques to predict the position of the missing planet he felt must be responsible. He called it "Planet X" and led unsuccessful searches until shortly before his death in 1916.

Even after Lowell's death, the observatory he founded continued his search. In 1929, the task of observing was given over to Clyde Tombaugh, a young researcher from Kansas who had just begun work at the observatory. Tombaugh began a systematic search of the skies for Planet X, and within a year had discovered a bright, distant object at about the position Lowell had predicted for Planet X. After receiving suggestions for its name from the public, it was decided to name the newly found object Pluto.

Tombaugh and other astronomers at Lowell Observatory continued their search through 1943, in the end covered roughly two-thirds of the sky—everything visible from their site in Flagstaff, Arizona. While Tombaugh discovered more than 700 asteroids in the course of the search, Pluto was the only outer solar system object discovered by the Lowell survey, which seemed to confirm its status as a true planet.

As scientists learned more about Pluto, however, it seemed less like the large planet it was originally thought to be. As discussed in later chapters, astronomers were able to rule out a large atmosphere for Pluto, which was puzzling. Other calculations showed that it should be much brighter than it is if it were Earth-sized (or even Mars-sized). As touched upon in other chapters, it was recognized that Pluto's gravity could not be responsible for the discrepancies in Neptune's orbit, and further planet searches were made. Eventually, with the *Voyager* flybys of Uranus and Neptune in the 1980s, more precise values for the masses of these planets became available and knowledge of their orbits were improved. Ironically, using the updated masses of Uranus and Neptune in orbital calculations removed any discrepancies in the orbit of Neptune—Lowell and other mathematicians were looking for a planet that didn't exist, and any similarity between Lowell's predicted position for Planet X and the location where Tombaugh found Pluto was coincidence and luck!

Meanwhile, other astronomers began to expect other objects should be found in the outer solar system. Gerard Kuiper theorized that a second asteroid belt might exist beyond Neptune, a belt of objects often called the **Kuiper belt objects (KBOs)** in his honor. After years of searching, the first Kuiper belt object was found in 1992. Hundreds of transneptunian objects (TNOs) have been found since 1992, some of which are in the Kuiper belt, others of which have different orbits, some of which are exceedingly similar to Pluto. This work led directly to the "demotion" of Pluto by the IAU in 2006.

Small Bodies in Popular Culture

The small bodies of the solar system have ingrained themselves in the popular imagination. Representations of comets have been found in the art of preliterate societies, which, given their interpretation as omens, is not surprising. Comets remained striking emblems through the Dark Ages and Renaissance; Giotto used a depiction of Comet Halley as the Star of Bethlehem in a painting in the early 1300s (see Figure 2.3), and the European Space Agency named their mission to Comet Halley after him. In the following centuries, comets have been used as the name for countless sports teams and household products, along with Bill Haley's backing band on "Rock Around the Clock."

Figure 2.3 The appearance of Comet Halley in European skies in the early 1300s inspired the Italian painter Giotto to include it as a representation of the Star of Bethlehem. Nearly 700 years later, the European Space Agency recognized Giotto's cometary art by naming their first mission to Comet Halley after him. Scala/Art Resource, New York.

The asteroids had a much later start on entering the public consciousness than did comets. By the turn of the twentieth century, however, visits to asteroids became commonplace in the newly evolving literature of science fiction. The asteroids served metaphorically as the setting for *The Little Prince*, though there was no attempt to be scientifically accurate by author Antoine de Saint-Exupéry. As authors recognized the potential of asteroids for both colonization and mining, asteroid colonies became settings for stories inspired by the Gold Rush and the old American West. On the other hand, the potential hazard they presented as impactors inspired books and movies about the aftermath of collisions and how those collisions could be avoided. Misconceptions about the density of the asteroid belt also led to dramatization of the danger of a ship passing through, perhaps most notably in the movie *The Empire Strikes Back*, though also appearing in more abstract form in the hugely successful video game *Asteroids*.

The most popular of the small bodies, however, is most certainly Pluto. Its discovery inspired the name of a popular cartoon dog, which has helped its popularity, but its status as the furthest planet has certainly been a factor. Perhaps oddly, the threat to its planetary status also added to its popularity, and in fact that popularity was used by some as an argument to maintain its status as a planet.

SURVEYS AND THE DIGITAL AGE

The last decades of the twentieth century found our society experiencing an information revolution due to the ready availability of computers. This revolution also had great influence in astronomy and small bodies studies in particular. The introduction of charge-coupled devices (CCDs) was a major step forward for observations. CCDs are an integral part of the digital cameras available today, and allowed quantitative measurements to be made in a way that was difficult or impossible with film photographs. Because CCDs store digital data, precise brightnesses can be measured. In addition, digital pictures never degrade with time or poor storage. Telescopes with CCD cameras can observe much fainter objects than was possible with film, extending the productivity of many observatories and small telescopes previously thought to be obsolete or nearing the end of their useful lives.

The availability of computers and digital data has also allowed automatic discovery of asteroids and comets. Development of software for automatic detection and discovery was spurred by and designed for the NEO surveys that began work in the 1990s, and it was a critical factor in the skyrocketing discovery rate that followed. However, such software has also been used by projects whose primary goal is not NEO discovery. The creation of CCDs with ever-larger areas has made all-sky surveys possible for very faint objects. Several of these surveys have been undertaken at various wavelengths, including the Sloan Digital Sky Survey (SDSS) and the 2MASS survey from the ground, and the Infrared Astronomical Satellite (IRAS) survey from space. These surveys were designed for studies of galaxies, quasars, and other distant phenomena, but as astrophysicists of the nineteenth century found, the survey fields were littered with asteroids. Again, because of the availability of digital data and computers, the information about asteroids from these surveys was easily extracted and has been a boon to planetary astronomers.

The computational power available to theorists and modelers also has allowed great strides to be made in simulating the formation of the solar system, and the dynamics and orbital evolution of objects. Furthermore, many of the calculations necessary for studying the possible effects of an impact into the Earth or another planet would have been impossible to do without the computers of today. The same is also true of image processing, with the techniques used in planetary geology often applied to software developed to look at medical imaging or even to touch up photos taken for advertising purposes.

STUDIES FROM SPACE

The final recent development in small bodies studies we will discuss here is the dispatch of spacecraft. The focus of the American and Soviet space programs on the Moon and planets (in particular Mars and Venus) still allowed some small bodies to be visited, usually as part of other missions. In this way the *Galileo* spacecraft performed the first visit to an asteroid, flying by 957 Gaspra and then 243 Ida en route to Jupiter. The first visits to comets had been performed in the 1980s by a flotilla of spacecraft to Comet Halley and Comet Giacobini-Zinner. While American and Soviet missions were frequently flown, there have also been dedicated European and Japanese missions to small bodies. Indeed, the Japanese *Hayabusa* mission was the first intended sample return from an asteroid, and the European *Rosetta* mission is en route to being the first lander on a comet. Other nations are also beginning involvement in small bodies missions as they can—Canada is pursuing a mission to discover and track NEOs from space.

The data returned from space missions has been a boon to planetary scientists. Close-up views of asteroid and comet surfaces have allowed us to gain a better understanding of the processes that are occurring there, which has translated into better knowledge of how they formed and how any hazard they might pose could be addressed. The samples returned from the *Stardust* spacecraft have already shown that our ideas about solar system formation are in need of revision, even as only a tiny fraction of the samples have been studied.

SUMMARY

The study of asteroids, comets, and dwarf planets has been central to the history of astronomy. From the pre-telescopic era when observations of comets showed that the heavens underwent changes, to spurring the improvement of orbital calculations in the 1700s, to the modern era where scientists have recognized the critical role of asteroids and comets in the extinction of species, small bodies research has contributed important ideas and important results both for other astronomical fields and for culture as a whole.

WEB SITES

The information on this Web site recounts how comets were seen as ill portents: http://www.bbc.co.uk/dna/h2g2/A3086101.

This site has a more general history of cometary studies: http://www.vigyanprasar.gov.in/dream/mar2001/comets.htm.

The history of asteroid studies is detailed at this Web site: http://dawn.jpl.nasa.gov/DawnCommunity/flashbacks/fb_06.asp.

This Web site focuses on the demotion of Ceres, Pallas, Juno, and Vesta from planet to asteroid: http://aa.usno.navy.mil/faq/docs/minorplanets.php.

3

The Orbits and Dynamics of Small Bodies

This chapter will discuss the places in the solar system where populations of small bodies can be found. Some of those populations are still speculative, with varying amounts of evidence for their existence. Other populations are unquestionably real. To understand the places that asteroids and comets are found, we will need to understand how to characterize their paths around the Sun, and the forces that affect them.

ORBITS

The paths followed by small bodies as they travel around the Sun are called **orbits**. In the 1600s, Johannes Kepler showed that objects orbiting the Sun travel in elliptical paths, and move most quickly when they are closest to the Sun and more slowly the farther away from the Sun they are. In addition, he showed that the time it takes an object to travel around the Sun (or its **orbital period**) was related in a predictable way to its average distance from the Sun (or its semi-major axis length). The semi-major axis is one of six **orbital elements** that are used to fully describe and uniquely distinguish an orbit. Two other important orbital elements for our purposes are the **eccentricity**, or how noncircular the orbit is, and **inclination**, or how the plane of an object's orbit differs from the plane of the Earth's orbit. The plane of the Earth's orbit is also called the **ecliptic plane**. The point in an orbit closest to the Sun is called the **perihelion** (also called q), the farthest point the **aphelion** (also called Q). The positions of these points relative to the Sun can be calculated as follows:

$$q = a(1-e)$$
$$Q = a(1+e)$$

where a is the semi-major axis of the orbit and e is the orbit's eccentricity. Semi-major axis is usually measured in astronomical units or AU, where 1 AU is equal to the mean distance from the Earth to the Sun. Inclination is measured in degrees; eccentricity has no units.

Isaac Newton generalized the findings of Kepler and solved the equations of motion for any two objects moving only under the influence of each other's gravity. More complicated arrangements, such as for three or more bodies, do not have simple solutions and, therefore, require numerical approximations. Newton found that all matter will exert some gravity, and that every object has a pull on every other object in the universe. This pull depends strongly on the mass of the object, and even more strongly on the distance to the object. For most objects in the solar system, the pull of the Sun dwarfs the pull from every other object, and everything other than the Sun can be ignored. For planetary satellites, the parent planet has a much stronger pull than the Sun, and the Sun is usually ignored.

RESONANCES

After the Sun, the most massive object in the solar system is Jupiter. Jupiter's gravity has had a profound effect on the orbital distribution of the asteroids and comets throughout solar system history. The other planets can also affect small body orbits. There are two main ways in which the planets change the orbits of comets and asteroids: close **encounters** and **resonances**.

Close encounters are relatively straightforward. Under normal circumstances, an object will orbit the Sun, with the gravitational force of everything other than the Sun unimportant. If a body happens to pass very close to a planet (say, Jupiter, for instance), that planet's gravitational pull on the body can become comparable in size to the pull of the Sun. The path of the body would then be influenced by both the Sun and the planet. The closer to the planet the body passes, the more its orbit can change. This type of orbit change is often purposefully used by spacecraft designers, for instance the MESSENGER mission used two encounters with Venus to direct that spacecraft to Mercury, and the *Voyager 2* mission took advantage of a rare alignment of the planets to be deflected by Jupiter to Saturn, by Saturn to Uranus, and then by Uranus to Neptune. While Jupiter is most effective at changing small body orbits, any planet can do it with a close enough pass. The asteroid 99942 Apophis will pass very close to the Earth in 2029, and the Earth's gravity will significantly change that asteroid's orbit. If it were possible to make perfectly accurate measurements, we would find that the

Earth's orbit also changed as a result of the encounter, ever so slightly. This is because the energy taken to change the orbit of Apophis around the Sun came from the orbit of the Earth around the Sun. However, because the Earth is so much more massive than Apophis, the change in Earth's orbit is not measurable.

The gravity of a planet can have much longer-term effects than during brief close encounters. If one graphed the orbital period of all of the known asteroids, there would be a distinct lack of objects with periods of 5.93 years and 3.95 years, though many other bodies have similar periods. The missing periods are related to the period of Jupiter, 11.86 years. If an asteroid had a period of 5.93 years, its closest approach to Jupiter would always occur at the same point in its orbit. Therefore, an extra force due to the gravity of Jupiter will pull the asteroid in a particular direction. This steadily changes the asteroid's orbit, increasing its eccentricity. An object with a period of 3.95 years would approach Jupiter at only two points in its orbit. The effect is not as strong on such an asteroid as one with a period of 5.93 years, but will still have a large effect given the time available since the solar system formed. This effect, called a mean-motion resonance, is important for many solar system bodies. These two resonances are called the 2:1 (Jupiter's period is twice that of the hypothetical asteroid) and 3:1 (Jupiter's period is three times the asteroid's). In the cases described here, the orbit of an object is *changed* under the influence of a planet. Some mean-motion resonances can, conversely, force an object to *stay* in an unusual orbit or serve as protection of a sort. For example, Pluto is in a 2:3 resonance with Neptune (it completes two orbits in the time it takes Neptune to complete three). This means that it always has its closest approach to Neptune in the same part of its orbit. However, Pluto's orbit is quite eccentric, and this close approach happens in the part of the orbit where Pluto is rather far from the Sun and Neptune. Pluto's orbit is so eccentric that it is sometimes closer to the Sun than Neptune, and therefore their orbits cross. However, at that time, Neptune is always far from Pluto. As a result of this resonance, Pluto and Neptune are never particularly close together, and Pluto's orbit is in no danger.

Another, more complicated type of resonance is called a **secular resonance**. In a secular resonance, the precession of two bodies occurs in tandem. This type of resonance is quite important for asteroids, and like the mean-motion resonances, secular resonances have led to the ejection of many objects from the main asteroid belt. The ν6 resonance with Saturn in particular has a major effect on the asteroidal population.

NONGRAVITATIONAL FORCES

The force of gravity alone is sufficient to explain and calculate the motions of most of the bodies in the current solar system (and, indeed, the universe). For small solar system bodies, however, nongravitational forces can become important and must be taken into account when predicting their motions over long periods. Ultimately, these other forces all derive their power from the Sun, though sometimes in indirect ways.

The **tails** and **comae** that distinguish comets from other objects carry gas and dust from the comet's **nucleus** (see Chapter 10). They also carry momentum from the nucleus, which leads to a change in the comet's orbit, much like a spacecraft orbit is changed when thrusters are fired. As the comet approaches the Sun and heats up, the rate of ice **sublimation** increases, as does the amount of nongravitational force on the comet. The exact change is dependent upon factors such as nucleus temperature, the composition of the ices in the nucleus, and the spin rate of the nucleus, among others. This makes it hard to predict the amount of nongravitational force that a particular comet will experience. However, measuring the difference between a comet's real orbit and the orbit expected with *no* nongravitational force can give a measure of the nongravitational force it has already experienced.

The Sun has a more straightforward connection with thermal forces. Matter takes time to heat up and cool down. This property is called **thermal inertia**. High thermal inertia material, like rock, takes longer to heat up and cool down than low thermal inertia material like sand. This property is part of the reason that the hottest part of the day, on Earth, is during the afternoon, rather than at noon—the ground takes time to heat up. The diurnal **Yarkovsky effect** is also due to this property. Light absorbed from the Sun carries a small amount of momentum with it. The force from this solar heating is symmetrical around the place directly beneath the Sun, where it is noon. Reradiated heat from the object is symmetrical around the hottest place on an object, where it is afternoon. The difference in these two directions puts a tiny torque on the object's orbit, changing it. Whether the orbit speed increases or decreases due to the Yarkovsky effect is dependent upon the direction of the object's rotation. This effect is very, very small, but for 1–10 km objects that have been going around the Sun for billions of years, it is measurable and important. Depending on the direction of rotation, this effect can either enlarge or shrink the object's orbit.

For objects with eccentric orbits, the seasonal Yarkovsky effect also occurs during the course of a year. In this case, the effect is because the warmest part of the year occurs at a different time than the time of the longest day (this is seen on the Earth, where the longest day in the northern hemisphere is in June, but the warmest part of the year is often in July or August). Here again, the directions of average incoming sunlight and outgoing thermal reradiation are not aligned, which changes the orbit. In the seasonal effect, the orbit always changes so that it gets smaller with time. As with the cometary nongravitational forces, the size of the effect is dependent upon specific properties of the object such as spin pole direction and spin rate, albedo or fraction of light reflected, thermal inertia, and size. The Yarkovsky effect was originally discovered in the early twentieth century, but was forgotten for decades before its rediscovery in the 1990s. It has been measured by precise radar observations of the asteroid 6489 Golevka, which moved 15 km due to this force over the course of 12 years.

Once objects are small enough, heat conducts efficiently and rapidly through them, so they are the same temperature throughout (or **isothermal**). The Yarkovsky effect does not occur on isothermal bodies. However, at this size range other forces can become important. **Poynting-Robertson drag** is similar to the seasonal Yarkovsky effect, but occurs for dust-sized, isothermal objects even if they are in circular orbits. In Poynting-Robertson drag, dust absorbs solar radiation and reradiates it. Einstein's laws of relativity show that the reradiation from the point of view of the Sun is not equal in all directions. This, again, changes the dust grain's orbit, making it smaller and having the effect of spiraling the dust grain into the Sun. At even *smaller* sizes, **radiation pressure** becomes most important. At these sizes, roughly 1 μm or smaller, the momentum carried by light itself pushes the dust grains away from the Sun, ultimately pushing them out of the solar system altogether.

THE RESIDENCES OF SMALL SOLAR SYSTEM BODIES

Given the large number of small bodies, and the forces that can act to change their orbits, it is no surprise they are found throughout the solar system. The following reservoirs are where most small bodies are found, or are hypothesized. We begin far from the Sun and travel inward.

Oort Cloud

At a distance of tens of thousands of AU, a good distance to the nearest star, we expect to find a large spherical shell of icy objects. At these distances, the Sun is only as bright as Mars or Jupiter appear at their brightest in the Earth's sky, and it can take upward of a million Earth years to complete an orbit. The Sun's gravity is quite feeble that far away, and it does not take much of a perturbation—a passing star, perhaps, or a close pass between bodies, to rip one of these objects from the Sun's grip altogether, or send it hurtling inward, after which interactions with the planets may make it a more regular visitor. This shell, commonly called the **Oort cloud** after astronomer Jan Oort, has never been directly seen, but astronomers are confident it exists. Why is this?

Oort and other astronomers noticed a pattern when looking at the orbits of comets: comets with orbital periods of roughly 200 years or less have orbits close to the plane of the planets; and comets with longer periods have orbits that can be very far from the ecliptic plane, and come from random directions. In addition, calculations of the **dynamical lifetime** of long period comets, that is, the typical length of time before a long period comet gets removed from the population either by striking an object or being perturbed out of the solar system, show that long period comets only last for a few tens of millions of years, compared to the 4.5 billion years that the solar system has existed. This implies a source where objects are still becoming long period comets. The fact that they have random inclinations suggests they are in a sphere centered on the Sun, and the size of the semi-major axes of their orbits gives an idea of how far that sphere is.

It is believed that there as many as a trillion objects in the Oort cloud, beyond 20,000 AU, with diameters of a few km. Objects still in the Oort cloud are too faint to be seen, so this number is very, very uncertain. We also do not know where the inner edge of the Oort cloud is, since objects closer in than ~20,000 AU are more difficult to perturb by passing stars, and as a result they end up in the inner solar system less frequently. Without their orbits to trace back, the existence of the "Inner Oort cloud" is still a matter of debate. Ironically, we know more about the structure of the region farthest from the Sun. Simulations suggest, however, that the Oort cloud does not extend closer than a few thousand AU. The known contents of the vast space between a few thousand AU and a few hundred AU are only a few objects, though it could be much more. The most notable object is 90377 Sedna, which does not come closer to the Sun than 76 AU, but retreats to nearly 1000 AU. There is disagreement over whether Sedna is an Oort cloud object that has been perturbed inward or an object from the Kuiper belt (see following section) that has been perturbed outward.

Transneptunian Region

The next grouping is the transneptunian objects, or TNOs. The region they inhabit comprises the space from the distance of Neptune (roughly 30 AU) to a few hundred AU. The TNOs are icy bodies beyond Neptune, and are thought to be similar to comets. Indeed, the short period comets (see following information) are thought to originate as objects from this region.

The TNOs are divided into several categories, dependent upon their orbits. A large number of them are found beyond Neptune in orbits near the ecliptic plane. This collection of objects is called the Kuiper belt (sometimes the "Edgeworth-Kuiper belt"), named after the Dutch-American astronomer who proposed its existence. The first Kuiper belt objects were found in 1992; hundreds are now known. Calculations suggest that roughly 70,000 objects larger than 100 km in diameter are present in the Kuiper belt. Interestingly, there appears to be a drop-off in the number of objects with semi-major axes more than 45–50 AU. This "Kuiper cliff" is still being investigated, but all indications are that there is a real lack of distant KBOs and not an observational effect due to the increasing faintness of those objects.

Another group of TNOs are found in resonant orbits with Neptune. Of these, most are in the 2:3 resonance, orbiting two times for every three Neptune orbits. As mentioned before, the dwarf planet Pluto is among these objects. For this reason, some call the smaller objects in the 2:3 resonance "**plutinos**," though this name is not an official one. Other objects are found in the 1:2 resonance with Neptune, and there are a few objects known to be in a 1:1 resonance (also called a Trojan resonance, see following section) with Neptune. All of these objects are currently thought to have been captured into resonant orbits early in solar system history.

Finally, some TNOs are in orbits that suggest they have had close encounters with planets, which have affected their paths. These can have high inclinations and large eccentricities, but are not in resonances. These bodies are called **scattered-disk objects**, and include the dwarf planet Eris.

Centaurs

Another group thought to be related to TNOs but orbiting closer to the Sun are the **centaurs**. These objects have orbits that cross the orbits of the giant planets, but are not in resonance with them. It has been suggested that the centaurs are similar to the scattered-disk objects, but they have simply been perturbed inward instead of outward. Some, including 2060 Chiron, the first one discovered, have shown evidence of a coma. This has resulted in some centaurs being listed as both comets and asteroids. Because their orbits cross those of planets, their dynamical lifetimes are short. While most of the objects discussed so far have likely been in their orbits for billions of years, the current centaur population is transient, and they have likely only been in their orbits for a few million years—only a few tenths of a percent as long as the other populations. Numerical calculations also suggest they will only last about that much longer before hitting a planet or being flung from the solar system altogether, while new objects replenish the population.

This part of the solar system is where Comet Halley spends most of its time. Its orbit suggests that it was originally from the Oort cloud rather than the Kuiper belt, but Comet Halley's orbit has been changed by close encounters with the outer planets so that it no longer retreats as far from the Sun at aphelion. Other comets with similar orbits are called **Halley family** comets. A distinction is also made for comets with periods longer than 200 years versus those with shorter periods. Not surprisingly, those objects that take longer than 200 years to orbit the Sun are called **long period** comets, while the others are called **short period** comets. Long period comets have semi-major axes beyond Neptune's orbit, while the short period comets have semi-major axes near or within Neptune's orbit. While the Halley family comets have semi-major axes that place them in the outer solar system most of the time, their perihelia are in the inner solar system, often interior to the Earth's orbit.

It is useful, in classifying comets, to use the **Tisserand parameter**, a number that describes an orbit. This number is calculated differently with respect to each planet. For Jupiter, it can be calculated as:

$$Tj = a_j/a + 2\sqrt{((a/a_j)(1-e^2))}\,cos(i)$$

Where *a, e* and *i* are the semi-major axis, eccentricity and inclination of the small body's orbit, respectively, and *aj* is Jupiter's semi-major axis (5.2 AU).

The Tisserand parameter doesn't change even after a close passage by Jupiter. If an encounter with Jupiter changes an object's semi-major axis, its eccentricity will also change so that *Tj* stays the same. Halley family comets typically have Tisserand parameters of less than 2 and periods of less than 200 years.

Trojans

The orbit of Jupiter begins the realm of the asteroids. The first group of asteroids we encounter as we move toward the Sun orbit at the same distance as Jupiter (5.2 AU). Thus, these objects are in a 1:1 resonance with Jupiter. Objects either leading or trailing Jupiter (or any planet) by 60° on average can remain there without being perturbed. This was first determined to be true by Joseph Lagrange, a French physicist, in 1772. In his honor, these points (as well as three other equilibrium points) are called Lagrangian points. The asteroids at the leading and trailing Lagrangian points of Jupiter (also called L4 and L5) are called the Trojan asteroids. Asteroids discovered and determined to be members of this group are typically named after participants in the Trojan Wars. Trojan asteroids do not need to be exactly at the 60° orbit difference from the planet, but from the point of view of Jupiter they move in paths (or librate) around the Lagrange points.

The population of objects in the Trojan region is thought to be quite large. The number of objects larger than 1 km is estimated to be several hundred thousand, with a total mass about 10,000 times less than the Earth. There is not much known about the Trojan asteroids besides their orbits. As will be discussed elsewhere, details of their origin and compositions are still under debate. However, it is thought that the Trojans are transitional objects, somewhere between asteroid and comet. Trojan asteroids can have very high inclinations compared to a typical main-belt asteroid. We do not know of any Trojan-type asteroids around planets other than Jupiter, Mars, or Neptune. Dynamical modeling suggests that Trojan asteroids of Saturn and Uranus would not be stable for long periods of time. Any possible Earth Trojans would be very difficult to detect from the ground, and the searches that have been done to date have been negative. Venus and Mercury Trojans are also thought to be dynamically unlikely.

Between the Trojan asteroids and the main asteroid belt are the **Hilda Group**, at the 2:3 resonance with Jupiter. The Hildas can be seen as having

Table 3.1. Major Asteroid/Comet Groups

Group	a (AU)	Stable Orbit?	Known Number	Predicted Number > 1 km
Oort cloud	>10000	Yes	0	>1 trillion?
Kuiper belt	30–50	Yes, in general	>800	100 million–10 billion?
HFC	~15–35	No	~50	?
Centaurs	10–30	No	~100	~10 million
Trojans	5.2	Yes	>2000	~300,000
Hilda	3.9	Yes	~1500	~25,000
JFC	2–7.5	No	>300	~1000–10,000
Main belt	1.9–3.4	Yes, in general	>300,000	~2 million
NEOs	0.9–2	No	4500	1100
Vulcanoids	0.09–0.2	?	0	Hundreds?

the same relationship with Jupiter as the Plutinos do to Neptune. It is not clear, however, whether the Hildas reached their current position from the Trojan region or from the main belt, or indeed, if they formed near their current orbits and are not derived from either group.

Jupiter family comets are those whose orbits are affected by Jupiter. They are typically defined as having *Tj* between 2 and 3. It is thought that Jupiter family comets originated in the Kuiper belt, as opposed to the Oort cloud origins of the Halley family comets. Again, while their semi-major axes place them near Jupiter most of the time, their perihelia can be near the Earth's orbit or even closer to the Sun.

Main Asteroid Belt

The **main asteroid belt** (often just called the "main belt" or "asteroid belt") spans the region between roughly 2 AU and 3.5 AU (sometimes the Hilda region is included in the main belt, sometimes not). Figure 3.1 shows the position of over 350,000 asteroids in terms of their semi-major axis and inclination. The region has noticeable concentrations of objects and areas with few or no objects at all. These latter regions, called the **Kirkwood gaps**, show the positions of some of the resonances mentioned before. The noticeable gap near 2.5 AU, vertical on Figure 1, is where objects would be in a 3:1 mean motion resonance with Jupiter. The ν6 secular resonance with Saturn defines the inner boundary of the main belt at inclinations below 15 degrees, aided by the 4:1 resonance with Jupiter near 2.06 AU. The 5:2 resonance with Jupiter is also visible near 2.8 AU. There is an obvious cutoff in the number of asteroids beyond 3.25 AU, near the position of the 2:1 resonance with Jupiter. Beyond that distance, the **Cybele group** orbit the Sun.

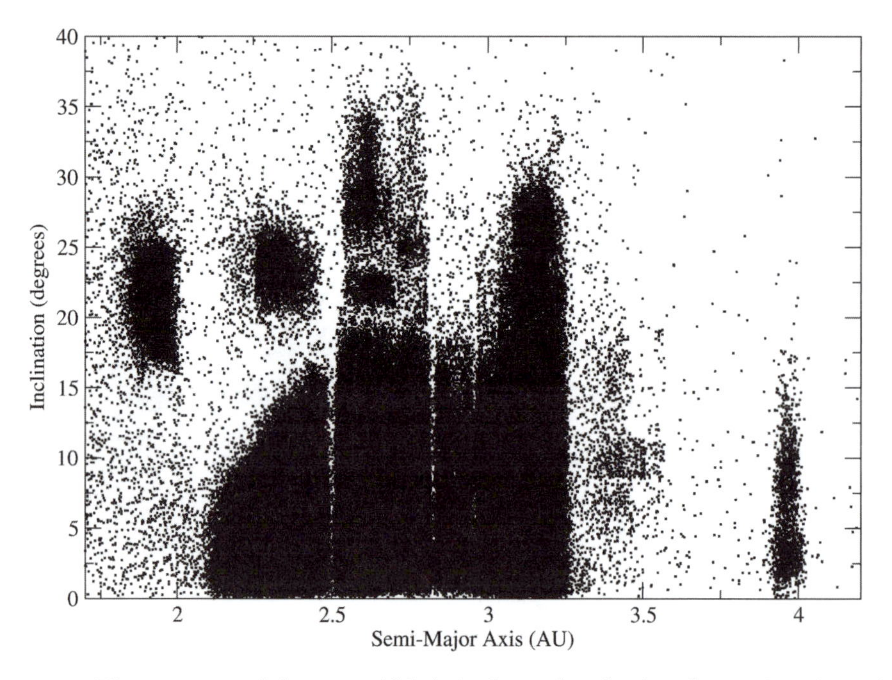

Figure 3.1 The structure of the asteroid belt is shown by plotting the semi-major axis and inclination of well over 300,000 asteroids. The vertical areas with few or no objects are resonances with the outer planets. The Hilda group is near 4 AU and separated from the main belt, as is the Hungaria group near 2 AU.

There are hundreds of thousands of known objects in the main belt. Roughly two million objects with diameters 1 km and larger are believed to be present in the asteroid belt. The size distribution is heavily weighted toward the small end, however; less than 5,000 of those objects are thought to be larger than 10 km.

Dynamical Families

Figure 3.2 shows a blowup of the middle of the asteroid belt. The Kirkwood gaps are still obvious and visible, but in addition, clusters of objects are obvious. These are dynamical families, first noticed by Japanese astronomer Kiyotsugu Hirayama in 1918. The Koronis family (from about 2.82–2.95 AU and with i of about 2°) and Themis family (i of 1–3° and semi-major axis from 3.1–3.25 AU) are two of the families identified by Hirayama shown in Figure 3.2. Families are named after their largest or earliest-discovered member.

Dynamical families originate via collisions in the main belt. Collisions can range from relatively minor events that create small amounts of **ejecta** (or fragments created in the impact), to large cratering events that can expel pieces of a meter in size or larger, to catastrophic impacts where the largest

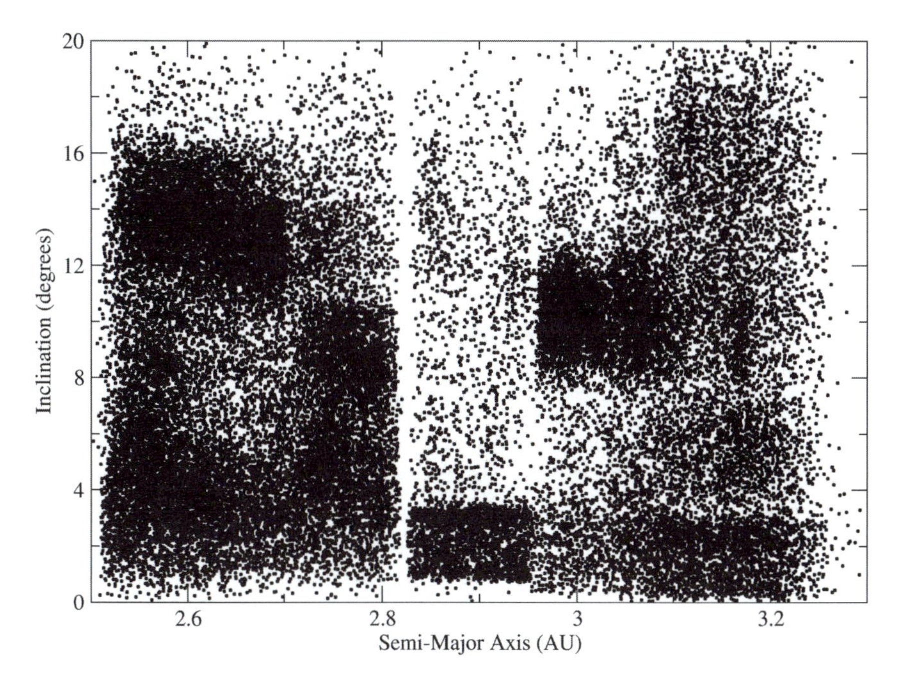

Figure 3.2 When investigated in detail, the asteroid belt shows high concentrations of objects with certain kinds of orbits. These clumps are dynamical families, objects on similar orbits that originated through impacts onto, and often the disruption of, larger bodies.

remaining intact piece is less than half the mass of the original target (the smaller object is always considered to be the impactor, the larger one the target). The exact result of an impact is critically dependent upon the speed and angle of impact, the relative sizes of the impactor and the target, and the composition and strength of the bodies involved. Because asteroids and comets are small, they do not retain most of the ejecta formed in an impact. This is in contrast to the Moon, Mars, Earth, or other larger objects.

The extra speed that ejecta have, compared to the target, leads to slightly different orbits. The differences between the speeds of the ejecta and parent bodies are small compared to their orbital speeds around the Sun, however, so the new orbits of the pieces are close to the original orbits. In some cases, like the Karin cluster (a subfamily within the Koronis family, where a fragment from that collision later suffered another collision), modeling has allowed the orbits to be calculated backward in time, resulting in an estimate of the time since the family forming event. For the Karin cluster, this is a short time—only between 5 and 6 million years. Most of the largest families—the Eos, Koronis, Themis, and Flora families, for instance—are thought to have originated billions of years ago, near the start of the solar system.

After a collision in the main belt, some of the new orbits for family members or ejecta may place them in a resonance. In addition, some of the newly created objects may be just the right size for the Yarkovsky effect to be a factor in its motion. While initially the orbits of family members are very similar, as time goes on the similarity can decrease and some objects can have orbits that are quite different from their initial orbit and the orbits of other family members. Another complicating factor is the existence of objects that coincidentally have orbits similar to members of a family. These objects, called interlopers, must be accounted for in detailed studies of families.

Dynamical families are also possible in other populations. The number of objects expected in the Kuiper belt is certainly sufficient to allow for collisions, although the collisional speed is much slower at those distances (impacts occur at roughly 1 km/s in the Kuiper belt compared to 5 km/s in the asteroid belt), and the number of known KBOs is much smaller than the number of known asteroids. Nevertheless, it is likely that families in the Kuiper belt will be identified in coming decades. There have been tentative identifications of Trojan asteroid dynamical families, as well.

Just interior to the main belt, the **Hungarias** can be seen at inclinations of 18–25 degrees and semi-major axes of 1.8–2.0 AU. These objects are separated from the main belt by a number of resonances, but are considered part of the main belt by some.

Planet Crossers

Mars has its aphelion at 1.666 AU from the Sun. Objects that have perihelia between 1.666 and 1.3 AU are called **Mars crossers**. These objects usually have aphelia in the main asteroid belt, and are thought to have reached their current orbits through a combination of weak resonances through the inner belt and the Yarkovsky effect slowly affecting them over a long period of time. The Mars crossers, like the centaurs, are not in orbits that are stable for long periods, and their orbits will continue to change quickly until they are removed from near-Mars space. In addition to these short-lived objects, Mars is the only terrestrial planet known to have Trojan-type objects. Less than 10 are currently known, 5261 Eureka being the largest with a diameter of a few km. Mars Trojans appear to have orbits that are potentially stable over the age of the solar system. Compositional studies, such as are detailed in other chapters, suggest the Mars Trojans were captured into their current orbits early in solar system history.

Small bodies with perihelia less than 1.3 AU from the Sun are classified as near-Earth objects, or NEOs. These include both asteroids and comets. These objects are of particular interest because some of them could potentially strike the Earth at some point in the future. Indeed, the Earth has been struck many times in the past, as described in more detail in the next chapter. On the other hand, their proximity to Earth also makes them relatively easy to visit with spacecraft compared to most other solar system objects. In fact, some NEOs are easier to visit by spacecraft than the Moon is.

Based on their orbits, NEOs are broken into further subclasses, shown in Figure 3.3:

1. **Amors**, whose semi-major axis is between that of the Earth and Mars, and whose closest approach to the Sun is outside the Earth's orbit.
2. **Apollos**, whose semi-major axis is larger than the Earth's, but whose closest approach to the Sun is inside the Earth's orbit.
3. **Atens**, whose semi-major axis is smaller than the Earth's, but whose farthest distance to the Sun is outside the Earth's orbit.
4. **Apoheles**, whose semi-major axis is smaller than the Earth's, and whose farthest distance to the Sun is inside the Earth's orbit.

Finally, there is another classification, that of **potentially hazardous asteroids** (PHA), which include all objects that have orbits that pass within 0.05 AU of the Earth's orbit.

The Aten and Apollo objects in particular are of interest because their orbits *cross* the Earth's, making these the most immediately dangerous bodies. However, the effects of close planetary encounters can change NEO orbits, sometimes drastically, and objects can evolve from one subcategory to another. Indeed, as discussed, all NEOs were originally from

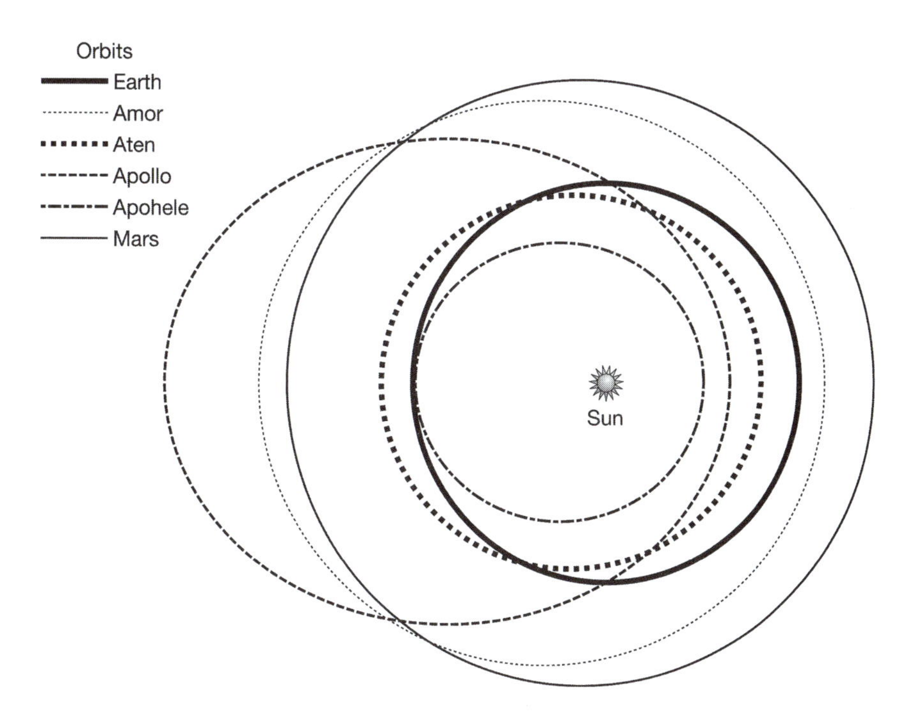

Figure 3.3 Near-Earth Objects are classified into different groups depending upon the specifics of their orbits. Amors and Apoheles do not cross the Earth's orbit but come close to it from the outside and inside, respectively. Apollos and Atens both cross the Earth's orbit but have semi-major axes larger or smaller than the Earth, respectively. The orbit of Mars is also shown, for comparison, along with typical orbits for these different NEO groups. Illustration by Jeff Dixon.

a different part of the solar system. The average time between becoming an NEO and either impacting a planet or the Sun is only a few million years, much shorter than the time since the solar system began. There is a constant delivery of objects from the main belt via resonances and the Yarkovsky effect to replenish the NEOs that are removed from the system.

The majority of known NEOs are in the Amor class. Very few Apoheles are known. It is currently thought that this is due to a bias in the way asteroids are observed and discovered. Because Amors are further from the Sun than the Earth is, there are times they can be observed all night long. Apoheles (and to a lesser extent Atens and Apollos) all spend some time, or even all of their time, in the daytime sky. This makes them very difficult to observe. As a result, careful modeling based on the best data available is required to come up with an accurate measure of how many Apoheles, Atens, and Apollos exist. Over 4,000 NEOs of all sizes are known, and it is believed that roughly 1,100 of them are larger than 1 km in diameter. About a quarter of all NEOs are also PHAs. The proportion of comets to asteroids in the NEO population is currently a matter of debate. Before the importance and efficiency of resonances was recognized, it was thought that nearly half of the NEO population had to be cometary to account for their numbers. More recent dynamical modeling in conjunction with compositional studies places the fraction of cometary NEOs closer to 15 percent at the most. The Tisserand parameter is often used to identify possible comet candidates in the NEO population.

Vulcanoids

Vulcanoids are a hypothetical population of small bodies with orbits near or interior to Mercury. They are named after Vulcan, a planet thought to exist in the 1800s and early 1900s. This planet was thought to be necessary to explain perturbations in the orbit of Mercury, in much the same way that Neptune's existence was predicted from perturbations of Uranus. Some observers even claimed to have seen transits of Vulcan across the solar disk. None of these observations were ever confirmed, however, and with Einstein's formulation of relativity, the need for an extra planet to explain Mercury's orbit disappeared.

Vulcanoids, if present, would represent leftover unaccreted bodies from the earliest solar system times. Searches using high-altitude aircraft and solar telescopes have found no objects interior to Mercury larger than 50 km, and the Yarkovsky effect would be expected to remove small objects efficiently. However, there is still a possibility that vulcanoids between these sizes could be found, because stable orbits are known to exist interior to Mercury down to 0.09 AU.

SUMMARY

There are four main areas where asteroids, comets, and dwarf planets are found: the main asteroid belt between Mars and Jupiter, the Trojan clouds leading and trailing Jupiter, the Kuiper belt beyond Neptune, and the Oort cloud stretching many thousands of times further from the Sun than the Earth's orbit. However, in addition to these stable orbits, small bodies can be found in orbits that are unstable over the span of tens of millions of years—these are the centaurs and near-Earth objects, among other smaller populations. The existence of objects in these unstable orbits means that bodies must move from stable regions to refill the unstable regions. It is thought that this occurs via impacts and the Yarkovsky effect for asteroids, while collisions and infrequent passages close to nearby stars change the orbits of comets.

WEB SITES

A flash-based online orbit simulator can be found at this Web site, including several preset situations (such as a Trojan asteroid-like case) and the ability to input positions and velocities to allow further exploration: http://phet.colorado.edu/sims/my-solar-system/my-solar-system.swf.

This site has a more technical introduction to the mathematics behind orbital calculations, including several further references: http://mysite.du.edu/~jcalvert/phys/orbits.htm.

The positions and orbits of comets, asteroids, and dwarf planets are provided by the NASA Jet Propulsion Laboratory at this Web site. Views are available of the orbits of individual objects, or positions of small bodies in large regions of the solar system: http://ssd.jpl.nasa.gov/?orbits.

The discovery of Sedna, thought to be the only Oort cloud object currently known, is detailed at this Web site: http://www.nasa.gov/vision/universe/solarsystem/planet_like_body.html.

For more information about further Hubble Space Telescope observations: http://hubblesite.org/newscenter/archive/releases/2004/14.

4

Meteors, Meteorites, and Meteoroids

The Earth is under constant bombardment from a rain of extraterrestrial material. In a typical year, the Earth is impacted by 54 tons of material. Most of it is the size of dust grains, and does not penetrate Earth's atmosphere. However, Earth is hit with objects the weight of a pencil or heavier roughly 100 times a day. Objects the size of marbles burn up in the atmosphere and are responsible for **meteors**, or "shooting stars." These can occur either randomly or in periodic, predictable meteor showers.

Chair-sized and table-sized objects often strike the Earth after spectacular fireballs, and fragments can survive to reach the ground as **meteorites**. Meteorite falls have occurred in Peekskill (New York), Monahans (Texas), and Tagish Lake (Canada) in the last 15 years. Earth bears ample evidence of even larger impacts. Every few decades, on average, Earth is impacted by an object of 10 tons or so—the mass of the Hubble Space Telescope. A house-sized piece of iron blasted a 1-kilometer-wide hole in the Arizona desert less than 50,000 years ago. Such an event occurs, on average, every few tens of thousands of years. An impact near the Washington Monument from an object of 300 m in diameter would leave a crater that would stretch from the Pentagon to the Capitol, with an ejecta blanket stretching halfway to Baltimore and destruction resulting across the Eastern United States.

Craters dot the North American landscape from Texas to Virginia, from Ontario to the Yucatan, as well as locations throughout the world. An impact in Mexico is thought to have resulted in the extinction of a large fraction of the Earth's life 65 million years ago, including the dinosaurs. It has been proposed that other so-called mass extinctions are also associated with impacts. Luckily, such large impacts appear to be relatively rare, with tens of millions of years between impacts of that size.

It is important to remember the difference between meteors and meteorites. Meteors all burn up in the atmosphere—one can never pick up a meteor. If an object from space makes it to the ground, it is called a meteorite. This is true regardless of the size of the pieces that land. These words can be easily confused. Indeed, the most famous landmark associated with an impact in the United States, Meteor Crater, is actually named incorrectly! A third concept with a similar name is that of **meteoroids**, small objects in space. Normally they are considered to be smaller than asteroids, but there is no official size range for asteroids or meteoroids. Before a meteorite strikes the ground, it could be called a meteoroid.

METEOR SHOWERS

Meteors can be seen on most nights under sufficiently dark skies. Often coming at a rate of a few per hour, and from random directions, these meteor phenomena are due to the chance encounters of small objects with Earth. These are called **sporadic meteors**. In contrast, several times a year **meteor showers** occur. In a meteor shower, the rate of meteors goes up from a few dozen to hundreds or even thousands per hour, and the paths the meteors follow seem to come from a particular spot in the sky, called the **radiant**. Meteor showers are named for the constellation in which their radiant falls; for instance, the Geminids appear to come from Gemini, and the Orionids from Orion. Meteor showers occur at the same time every year—every August 12th, the Perseids are at their peak, for example. However, the rate of meteors can vary substantially.

While the Perseids consistently provide about 100 meteors per hour near their peak, the Leonids have varied from years when few meteors were seen beyond the typical sporadic contribution, to years like 1833 (see Figure 4.1), where over 100,000 meteors per hour were seen. That level of activity, often called a **meteor storm** rather than a shower, translates to over 25 meteors per second. More recent Leonid storms occurred in 1966 and around the turn of the twenty-first century. These storms had a more modest rate of "only" a few thousand meteors per hour, but they were unforgettable for those who witnessed them.

The clouds of particles that cause meteor showers orbit the Sun, like everything else in the solar system. We know the position at a specific time (for instance, the Perseid meteors are at the Earth's position relative to the Sun on August 12th of every year), we know the direction of motion (out of the constellation Perseus, in this case), and can measure their speed. When put together, the characteristics of the orbit can be calculated. In the case of the

Figure 4.1 This depiction of the 1833 Leonid shower, though produced over 50 years later from secondhand accounts, gives a sense of the nature of meteor storms, with dozens of meteors visible at all times. This meteor storm gave rise to the first scientific explanation of meteors as extraterrestrial in origin rather than being simply weather-related. NASA.

Perseids, the calculated orbit is very similar to that of a comet, Swift-Tuttle. Indeed, in many cases meteor showers are associated with comets. It is for this reason that, in general, meteors are thought to be derived from comets.

The Discovery of Meteor Showers

It was not until the Leonid storms of the 1830s that meteors first received real scientific attention. Previously they had been thought to be atmospheric phenomena, and were named accordingly (the word *meteor* has the same root word as *meteorology*, the study of weather). American astronomer Denison Olmsted observed the 1833 storm from Connecticut and noted that all the meteors appeared to emerge from a discernable region of the sky, the radiant. He collected reports of observations from throughout North America, and first proposed that meteors were coming from outer space rather than from within Earth's atmosphere. It was not long before the annual occurrence of the Perseids was recognized, and other meteor showers also identified. (Interestingly, the Perseids were identified by nonscientists centuries earlier—the shower was known to German peasants as "the tears of St. Lawrence," since it fell on the anniversary of his death.)

METEORITES

Meteorite discoveries are divided into two categories, **falls** and **finds**. Falls are meteorites that are seen as fireballs and traced to their resting place. Very rarely, this can be done without much effort, such as the handful of meteorites that have struck houses (and even cars!) or fallen within the sight of people. More often, after reports of a fireball, a search area can be defined and fragments found. With video evidence, such as with the Tagish Lake fireball/meteorite, searches can be done quickly and effectively. Other meteorites are never found due to sketchy or contradictory eyewitness accounts of a fireball's speed and direction. Falls are scientifically very interesting because they have the best chance of retaining delicate minerals that are stable in the vacuum of space but quickly degrade with exposure to Earth's atmosphere.

Finds are meteorites that are found without evidence of a fireball. Unlike falls, where the amount of time spent on Earth is known to be very short, finds may have spent thousands of years or longer on Earth before being recognized as meteorites. Most meteorites in the world's collections were found in Antarctica, where the United States and Japan both send teams every year for the purpose of collecting finds. There are several reasons Antarctica has excellent meteorite hunting grounds. First, and perhaps, most obviously, Antarctica has vast stretches of ice where almost any rock encountered has a

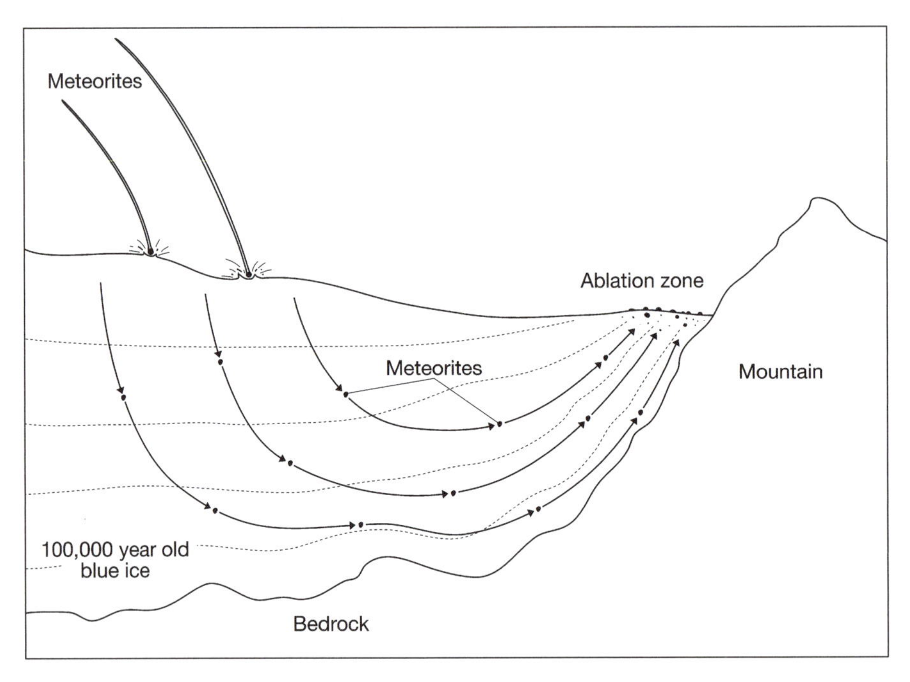

Figure 4.2 Meteorite searchers take advantage of a natural concentration mechanism to increase their odds of finding meteorites in Antarctica. By studying the flow of glaciers, certain areas have been identified where meteorites are carried long distances and then left behind where the glaciers melt. Expeditions searching these areas have had excellent success and typically bring back hundreds of meteorites per year. Illustration by Jeff Dixon.

high likelihood of being a meteorite. Searches center on particular areas where glaciers are known to be eroding at a rapid rate. Because glaciers move, and because the ice erodes rapidly but rock doesn't, any meteorites that might have fallen on the glacier are collected in one place.

Before the Antarctic expeditions began, most meteorites were found in areas where the wind is removing soil, leaving any ancient meteorites behind. The discoverers were often farmers, who were familiar with their land and recognized any unusual rock types. To this day, nomads in the Sahara desert find a significant number of meteorites for similar reasons. Because these unusual rocks were not always recognized as scientifically valuable, there are many stories of meteorites sitting on mantelpieces for decades, or even being used as doorstops!

Meteorites are typically named for the town nearest their fall or find (in populated areas, the nearest post office). For areas that have no nearby towns, geographic features are used (for instance, Tagish Lake or even Sahara). Antarctic meteorites come from a few specific areas and are numbered as well as named, with the number derived from the year of collection and the order in which it was characterized. Hence, ALH 84001 was collected in the Alan Hills of Antarctica in 1984 and was the first one to be characterized.

Recognizing Meteorites

Finding meteorites outside of a dedicated expedition is a very, very improbable event. However, it does happen a handful of times each year. Meteorites have several characteristics that are unusual compared to normal rocks: they have smooth sides and a coating called a fusion crust due to their passage through the atmosphere; they usually contain at least some metal; and they do not contain layers of any kind. There are resources on the Internet to help determine if any suspect rocks are meteorites or "meteor-wrongs." For those who prefer to leave the searching to others, some meteorites are common enough that they are sold at reasonable prices by reputable dealers.

CLASSIFICATION OF METEORITES

Meteorites are classified into different groups defined by their compositions. The first classification system was very general and included three groups: **stones, irons**, and **stony-irons**. This level of classification requires no special equipment, and requires no special training: stony meteorites appear to be rocks, iron meteorites are hunks of metal, and stony-irons contain both metal and rock. While most meteorites are thought to come from asteroids, we know that some meteorites originate on the Moon and Mars. Those meteorites will not be discussed in this book, but descriptions of them and what we learn from them can be found in other volumes.

The original object that the meteorites came from is called the **parent body**. Some parent bodies are thought to have been destroyed in a large impact, with the meteorites found being all that remains of the original. Other parent bodies are thought to still exist today (the Moon and Mars and large asteroids like Vesta or Ceres are the most obvious examples), with the meteorites being knocked off in relatively small collision events.

Through laboratory investigation of meteorites, scientists have found variation within all three of these groups, and meteorite scientists (or meteoriticists) use a different scheme, separating meteorites into two major categories: **chondrites** (all of which are stony) and **achondrites** (including all irons and stony-irons, as well as some stones).

CHONDRITES

Chondrites are so-called because they contain small, mostly spherical bits of glass called **chondrules**. The abundance of chondrules can range from a small amount all the way up to comprising most of the meteorite. The origin of chondrules is something of a mystery at present. They show evidence of having reached high temperatures (over 1500°C) in only a few minutes, cooling in the course of a few hours. It is not known what caused the heating, with leading candidates including shocks, lightning in the solar nebula, and outbursts from the early Sun, among other possibilities. Each candidate has proponents, though none can yet explain all of the characteristics seen in chondrules.

The part of a chondrite meteorite between the chondrules is called the matrix and is composed of very fine-grained minerals. The chondrites are thought to have survived largely unchanged since their formation roughly 4.5 billion years ago. When a mixture of minerals is heated to a given temperature and pressure, a new set of minerals is formed in equilibrium with one another, with compositions appropriate to the temperature and pressure that was reached. The combination of minerals found in chondrites is often **unequilibrated**, with no temperature and pressure conditions that can explain all of the compositions of the minerals. This shows they have not experienced high temperatures since the time these different minerals were brought together. A period of high temperatures would have led the minerals to exchange atoms and led the set of minerals to reach equilibrium with one another.

Chondrites are divided into three major classes, the **carbonaceous chondrites**, **enstatite chondrites**, and **ordinary chondrites**. (Ordinary chondrites are named so because they are very common among meteorites, not because they are uninteresting!) The ordinary chondrites are the meteorites most commonly seen to fall, amounting to roughly 80 percent of falls. The ordinary chondrites are further subdivided into the H, L, and LL groups depending on the amount of iron they contain—H has the most, L less, and LL the least. The H chondrites are the most common of the ordinary chondrites, accounting for nearly 40 percent of all falls. The L chondrites are somewhat less common, at roughly one-third of all falls, and the LL chondrites account for slightly less than 10 percent of all falls.

Carbonaceous chondrites were named because at the time they were thought to contain more carbon than other meteorites, though more recent studies have shown that, in fact, these meteorites do not always have more carbon than other types of meteorites. Today, the presence of specific elemental ratios is used to recognize members of this group. Like the ordinary chondrites, the carbonaceous chondrite group is subdivided into several subgroups: the CI, CM, CV, CO, and CR subgroups are the major divisions, differing in metal content and the minerals that are present. Roughly 5 percent of all falls are carbonaceous chondrites.

Two components of carbonaceous chondrites are worth particular notice and are further evidence of their ancient nature. Whitish-colored specks and grains are found in some carbonaceous chondrites. When analyzed, these grains turn out to be rich in calcium and aluminum, giving rise to the name **calcium-aluminum-rich inclusions**, or **CAIs**. The CAIs are the oldest known objects in the solar system, with ages of 4.57 billion years since formation, roughly 2 million years before the event that formed the chondrules.

There are other small grains whose atomic composition is very different from those seen in any other rocks, or from most of the grains in their host meteorite. These are called **presolar grains** and are believed to have formed in other stars and been ejected in supernova explosions or in outflows from red giants. These grains, according to theory, randomly found themselves near the Sun very early in solar system history. Most of them were heated and destroyed, becoming homogenized with the rest of the material that became the solar system, but a small number (luckily) escaped that fate and were incorporated into meteorites. We do not know enough about the grains or the candidate stars that may have been the original location of the presolar grains in meteorites, but it seems possible that rocks in future star systems could include grains that formed in our own Sun.

Unlike the carbonaceous chondrites, the name for the enstatite chondrites is an apt one, as these have a large fraction of the silicate mineral enstatite. Enstatite is a type of **pyroxene**, one of the most common mineral groups on the Earth. Enstatite contains little or no iron and is very rarely found on the Earth's surface, but is predicted to be rather more common in the Earth's mantle. It has also been identified in disks of dust orbiting other stars. Geochemists have determined that enstatite chondrites must have formed in water-poor and oxygen-poor parts of the solar nebula, in contrast to the other chondrites, particularly the carbonaceous chondrites. The enstatite chondrite group accounts for less than 2 percent of all falls.

ACHONDRITES

Meteorites without chondrules are called achondrites. These meteorites have experienced a variety of heating and melting processes that have modified their compositions and original textures. The meteorites that have only experienced a little melting are called **primitive achondrites**. More commonly found are pieces of objects that have been heated enough to separate into a crust, mantle, and core, or **differentiated**.

Some iron meteorites are pieces of asteroidal cores, formed when their parent bodies differentiated. These meteorites, called **magmatic irons**, give insights into the conditions and composition at Earth's core, for which we have no samples and little hope of directly measuring. Brachinites are meteorites very rich in **olivine**, the most common mineral in the Earth's mantle (and called peridot when used as a precious stone). Olivine is detected throughout the solar system and beyond and is composed of very common elements: magnesium, iron, silicon, and oxygen. Because of the high concentration of olivine they have, brachinites are thought to be from asteroidal mantles. These are quite rare, perhaps because they are not strong enough to survive collisional evolution in the asteroid belt and passage through Earth's atmosphere.

There are also several kinds of meteorites that represent crustal material. The most common are the **HED meteorites**. This group (standing for "howardite, eucrite, diogenite") is thought to come from the asteroid Vesta, providing one of the few well-accepted connections between a meteorite and parent body. The connection was formed by both geochemical and astronomical arguments—the reflectance spectrum of Vesta was found to be a very close match to the spectra of HED meteorites obtained in the laboratory (for further details of this technique, see Chapter 7). Independently, geochemists studying the HED meteorites concluded they had to originate from a large object that was likely still intact, and Vesta was the best candidate. Further work studying main-belt and near-Earth objects sharing Vesta's spectral and orbital characteristics has strengthened these conclusions.

Another group of melted meteorites are the angrites. Laboratory experiments and theoretical calculations have been performed on the angrites and HED meteorites (Vesta) to try and determine their compositions before they were heated and melted. The results suggest that the parent bodies of both of these meteorite groups had carbonaceous chondrite-like compositions before they were melted.

Unlike the achondrites mentioned before, there are several achondrite groups whose origins are not obvious and are still a matter of controversy. The **non-magmatic irons** are thought to have formed through localized impact melting of chondritic materials rather than via melting of the whole object and differentiation. The stony-irons have two main groups, the **mesosiderites** and **pallasites**. Both are roughly equal mixtures of silicates and metal, with the pallasites often containing large crystals of olivine. Indeed, the name of the mesosiderites reflects their general composition (*meso* meaning intermediate or middle, while *siderite* is an older term for iron meteorites). Pallasites, confusingly, have nothing to do with the asteroid Pallas, but were named after the German naturalist Peter Pallas years before the discovery of the asteroid that unrelatedly shares his name.

The formation of stony-iron meteorites is not well understood, but there is general agreement that it was a complicated process involving melting and differentiation followed by impacts with other objects while still molten.

INTERPLANETARY DUST PARTICLES

The smallest meteorites usually don't even make it to the ground. As mentioned previously, most of the mass that hits the Earth is in the form of dust. Regardless of how dedicated or fastidious the search team, these **interplanetary dust particles** (or **IDPs**) will never be seen to fall or collected in the usual way. Instead, collection is done by airplanes or balloons in the stratosphere. Chemical analysis of IDPs is difficult, but they have been roughly separated into at least two groups, **hydrous** and **anhydrous**, depending on whether or not they contain water in their minerals. The hydrous IDPs are similar to the CI carbonaceous chondrites, while the anhydrous IDPs are not similar to any meteorite type. This has led many scientists to suspect that the anhydrous IDPs are cometary in origin, while the hydrous ones are thought to be formed as the result of collisions between asteroids. In addition, there are some IDPs that appear to be from other star systems, similar to the presolar grains in carbonaceous chondrites.

The *Ulysses* spacecraft, a mission designed to study the Sun and solar wind, found streams of dust moving through the solar system, and based on the direction they were moving, scientists believe some of those streams come from outside the solar system. The *Stardust* spacecraft, whose main mission was to return samples of dust from a comet tail, flew through one of those streams hoping to capture and return some samples to Earth, a study whose results will become known in coming years. Further information about *Stardust* and small bodies missions in general can be found in Chapter 12.

COMPOSITION OF METEORITES

If one could count every atom in the Sun, we would find that some elements are more numerous than others. Hydrogen is by far the most common element in the Sun, followed by helium. Every element down to uranium would be found, however, to some degree. The relative amounts of all of the atoms in the Sun are called **solar abundances**, some of which are shown in Table 4.1.

When we look at the chondrites and measure the elements present, we find that most of the elements are present in the same ratios as in the Sun—the relative amounts of magnesium and silicon, for instance, or of ytterbium and praseodymium. However, some elements are rarely found in the chondrites compared to the amounts found in the Sun, like helium, simply because those elements do not remain easily bound to rock. The distributions of elements in the chondrites are, not surprisingly, called **chondritic abundances**, also shown in Table 4.1. The similarity between the chondritic and solar abundances is consistent with the expectation that the bodies of the solar system all started out made of the same components as the Sun, modified for different temperatures and gravity in different parts of the solar system. For instance, there is less nitrogen in the chondrites compared to the Sun because nitrogen does not commonly occur in rock-forming minerals. The major rock-forming minerals, like silicon, magnesium, iron, and aluminum, all occur in the same ratios in the chondrites and the Sun to within a factor of two, which is considered an excellent match in this type of analysis.

Table 4.1. The Relative Amounts of the Elements in Chondrites

Element	Abundance in Sun Relative to Silicon	Abundance in CI Chondrites Relative to Silicon
Hydrogen	28000	0.19
Carbon	10	0.32
Oxygen	24	4.6
Magnesium	1	0.93
Iron	0.9	1.8
Gold	0.0000003	0.0000013
Aluminum	0.082	0.082
Sodium	0.06	0.047
Nitrogen	3.1	0.03
Nickel	0.05	0.1

If we could count every atom that makes up Earth, we would expect to find it to have chondritic abundances of the elements. However, the elements would not be evenly distributed through Earth. Most of Earth's nickel, for instance, is found in the core and very little is in the mantle or crust. By contrast, most of Earth's magnesium is in the mantle and crust, with little in the core. Elements that tend to be found in the core are called **siderophile** (metal loving), while those found in the crust and mantle are called **lithophile** (rock loving).

ISOTOPES AND WHAT THEY TELL US

As you are probably aware, atoms consist of protons, neutrons, and electrons. Elements are defined by the number of protons in their atoms—if an atom has one proton it is hydrogen, if it has two it is helium, if it has six it is carbon, and so on. However, atoms from the same element can have different numbers of neutrons, and thus different masses. These are called **isotopes**. Carbon has three isotopes, with atomic weights of 12, 13, and 14, symbolized like so: ^{12}C, ^{13}C, and ^{14}C.

Isotopic studies of meteorites are used in several ways. Many isotopes are radioactive, changing into stable isotopes of other elements after their decay. For instance, ^{14}C decays into ^{14}N. Geochemists can study the isotopic mix of elements in minerals and calculate how long it has been since they have formed. As an example, potassium (K) is an element commonly found in rocks. One of the potassium isotopes, ^{40}K, decays into argon (^{40}Ar) at a known rate. Argon does not form compounds with other elements, and is not found in minerals except as trapped gas molecules. So by measuring the amount of ^{40}Ar found in a rock, and measuring the amount and kinds of K isotopes in the same rock, the age of the rock can be calculated. It is on the basis of measurements like this using various elements that **radiometric ages** have been calculated for everything from pottery shards and archeological sites to meteorites, lunar samples, and the planets.

Stable isotopes have also been used to gain insights into meteorite histories. The mix of oxygen isotopes—^{16}O, ^{17}O, and ^{18}O—is thought to be fixed on an object since its formation. The relative amounts of those three isotopes are often used as a "fingerprint" of a parent body. Lunar meteorites were identified as such in part because they have the same oxygen isotopic ratios as samples returned by the Apollo astronauts. In turn, theories of the Moon's origin must account for the fact that the Moon and Earth share the same oxygen isotopic ratios and must have had a common origin. Different meteorite types are grouped together based on their isotopic ratios, and conversely those ratios can be used to conclude that similar-seeming meteorite groups formed on totally separate bodies.

Another important set of isotopes are those of hydrogen. Most hydrogen has a proton but no neutrons. Deuterium is a form of hydrogen with both a proton and a neutron. The hydrogen in the Earth's water is roughly 0.015 percent deuterium. This fraction, sometimes called the **D/H ratio**, is also used as a fingerprint. The D/H ratio in comets has been measured as significantly higher than that on the Earth, suggesting that cometary impacts may not have been an important source of the Earth's water, or at the least that water with a much lower D/H ratio must also have been brought from another source or present on the early Earth in order to have the final ratio we see today.

PROCESSES ON METEORITE PARENT BODIES

Meteorites provide a record of many different processes since the solar system began. While dramatic changes are rare, especially in the last 4 billion years, subtle change occurs to this day in the asteroid belt.

The most drastic changes seen in meteorites are the result of heating. As the meteorite parent bodies were being formed, they accreted varying amounts of radioactive elements. As those elements decayed, they gave off heat. Larger objects retained more of this heat than smaller objects, and those that formed more rapidly had a longer time to capture the radiogenic heat than those that formed slowly. As a result, some large objects got hot enough to melt. Starting with a chondritic composition, small amounts of melting would create the primitive achondrites. As temperatures continue to rise, the molten metal and silicates would begin to separate because liquid silicates did not have the strength to support the much denser metal. The separated metal would sink and pool with nearby metal, accelerating the process. The sequence would end with the metal near the center of the parent body as a core. Evidence from meteorites suggests that this happened in dozens of parent bodies, perhaps as many as 75–100. This evidence comes not only from the iron meteorites that obviously result, but also from the silicates that are left behind after differentiation.

There is also ample evidence in the meteorites for **thermal metamorphism**. This is heating to temperatures short of the melting point for the materials in the rock. As heating increases in chondrites, the minerals equilibrate with one another, followed by changes in the grain sizes of minerals in the matrix. In the final stages before melting, chondrules become indistinct and begin to merge with the matrix.

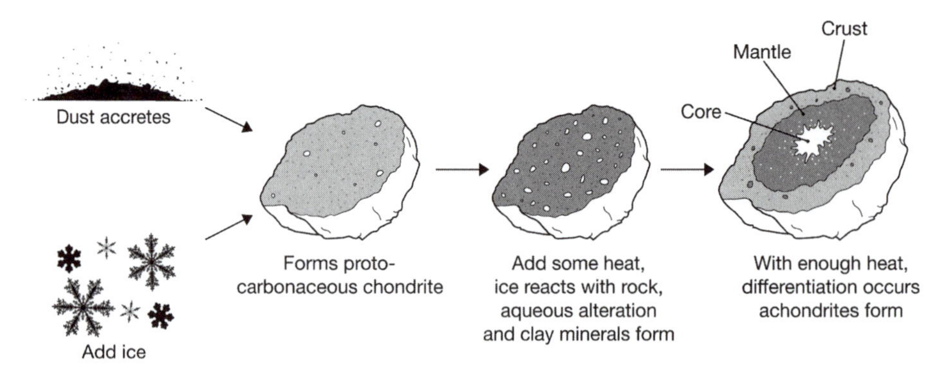

Figure 4.3 This schematic shows how meteoritic materials change with heat. In this case, a carbonaceous chondrite is formed from a mixture of rock and ice. With the addition of some heat, the ice melts and the water that is released reacts with the rock to aqueously alter the body. If enough heat is added, the parent body starts to melt and separate into crust, mantle, and core. Illustration by Jeff Dixon.

In addition to finding thermal metamorphism in chondrites, many carbonaceous chondrites (and a few other chondrites) also have been **aqueously altered**, with minerals that have had reactions with water. In some cases, water (or a part of water, like an OH ion) actually becomes part of the new minerals after the reaction. CI chondrites are almost entirely made up of aqueously altered minerals containing water or OH. Some meteorites (like some CV and CO meteorites) show evidence of once having these types of minerals, but then experiencing further heat and dehydration.

There are even some meteorites from which some fluids have been recovered. Interestingly, those meteorites are ordinary chondrites, whose parent bodies were not thought to contain much water. Taken as a whole, these findings suggest that water may be very common on chondrite parent bodies, and are suggestive to some that much of Earth's water may have arrived via meteorite impact.

Some processes continue to this day on meteorite parent bodies and can be seen in the meteorites that make it to the ground. Impact melting, mentioned previously, occurs as some of the kinetic energy of an impactor is converted into heat, via shocks. While this heat dissipates relatively quickly, it can be enough in some cases to melt areas close to the impact point. Often, meteorites with impact melt also contain unmelted fragments.

Rock that has been impacted can also experience changes, and those changes are seen in meteorites. In the simplest case, impacts break large rocks into smaller pieces. On parent body surfaces, this process eventually results in the creation of a loose powder, the **regolith**. However, unless regolith powder is held together somehow, it cannot survive the chain of events that results in a meteorite. **Regolith breccias** are the meteorites that result from the cementation of regolith, the process by which impact-generated shock and pressure makes solid rock out of the powdery regolith.

Impacts also create shock waves that propagate through the target. These shock waves create very high temperatures and pressures, but only for a very short time. As mentioned already, often some melting takes place. Even without melting, however, shocks damage minerals inside rocks, with results that are visible under the microscope. By conducting experiments with shocks of known pressures, the specific effects seen in particular meteorites can be converted to a value for the shock pressure that meteorite experienced. Interestingly, shocks from impacts are also thought to be responsible for cementation of regolith into regolith breccias.

In addition to these violent, sporadic events, meteorite parent bodies also experience gentler, but more constant processes. An example of such a process that can be seen in the meteorites is interaction with and capture of the solar wind. Along with photons, the Sun is emitting a steady stream of particles, mostly protons and electrons. In addition to those particles, however, roughly 5 percent of the total are elemental nuclei—mostly helium, but other elements as well. The solar wind particles can become implanted in parent body regoliths, and many such meteorites have been found, called gas-rich meteorites. Particles with enough momentum, whether from the solar wind or cosmic rays from beyond the solar system, can damage the crystalline structure of rock, as well. Because they cannot penetrate terribly far, finding tracks from the solar wind or cosmic rays in a sample implies that it spent time near the surface of its parent body. Knowing or estimating the amount of damage that would accumulate with time enables a rough age to be determined. The distribution of tracks can even be used to show which way the sample was oriented on its parent body.

SUMMARY

The Earth is constantly being hit by extraterrestrial material. Dust-sized material burns up in the atmosphere as meteors (or shooting stars), while larger pieces that survive passage to the Earth's surface are called meteorites. Meteorites are thought to come from asteroids (save those few that originate on the Moon or Mars), while meteors seem to come both from asteroids and from comets. Meteorites are classified based on their compositions, with most having abundances of the rock-forming elements similar to those found in the Sun. Laboratory studies of meteorites and cosmic dust are critical pieces in our understanding of the small bodies, allowing better understanding of telescopic observations. These samples have shown us that that some asteroids have experienced large-scale melting and volcanism, others have been affected by water and lesser amounts of heat, and some are virtually unchanged since the origin of the solar system.

WEB SITES

The Web pages for the Antarctic ANSMET expeditions can be found at this site. *Brother Astronomer: Adventures of a Vatican Scientist* by Guy Consolmagno includes an account of his experiences on an Antarctic meteorite expedition: http://geology.cwru.edu/~ansmet.

Web pages from Washington University with an abundance of material about meteorites: http://meteorites.wustl.edu.

Web pages from the University of Arizona with material about meteorites: http://meteorites.lpl.arizona.edu.

Sky and Telescope magazine includes a history of scientific study of meteors at this Web site: http://skytonight.com/observing/objects/meteors/3304116.html.

NASA's Web page for the *Genesis* mission, including discussion of isotopic studies: http://genesismission.jpl.nasa.gov/gm2/news/features/isotopes.htm.

5

The Formation of the Solar System and the Small Bodies

By studying the meteorite samples in our collections (as discussed in Chapter 4), observing the small bodies through a variety of methods, and visiting them with spacecraft, scientists have been able to craft a timeline for the formation of the small bodies. In combination with observations of the planets and other star systems, and of the Earth itself, we have the outlines for the formation of the solar system as a whole. In this chapter, we will consider how the solar system formed, and what secrets the small bodies hold from those early times.

CHARACTERISTICS OF THE SOLAR SYSTEM

On the face of it, there would seem to be little prospect of successfully describing the formation of the solar system since the distances are vast, and the time is much, much longer than a lifetime. However, human beings have been interested in the subject for thousands of years and have made untold observations, which have led to numerous explanations for what they have seen. We continue to be led by observations, which today are made by ever-more-sophisticated instruments. Any theory of the solar system's formation must explain the following observations, many of them centuries old:

1. There are several planets, which have different sizes.
2. All of the planets orbit the Sun in roughly the same plane.

3. The planets all have more-or-less circular orbits.
4. All of the planets orbit in the same direction.
5. Some planets are rocky, others are dominated by gas.

THE NEBULAR HYPOTHESIS

From observations of star-forming regions, along with theoretical modeling, astronomers believe that the Sun began its life in a giant molecular cloud (GMC), a dense region of space full of dust and gas. These GMCs can be quite large (50 light-years or more in size) and massive (up to hundreds of solar masses), and are composed mostly of hydrogen and helium, with a small amount of other molecules present. Such a cloud will naturally have some areas that are slightly denser than others and are moving with slightly different speeds. Given a disturbance like the shock wave generated by a nearby supernova, these denser areas can begin to collapse to become denser regions yet. A dense region like this is thought to have been the site where the Sun's formation occurred and is called the **solar nebula**. In general cases, it is called a **protoplanetary nebula** or **pre-stellar nebula**. The composition of the solar nebula was largely hydrogen and helium, with every other element adding up to less than 2 percent of the mass of the system. Of these elements, most of the hydrogen and some of the helium were created at the formation of the universe (popularly known as the Big Bang), while the rest of the helium and every other element formed inside of a star through nuclear fusion or supernova explosions, or via radioactive decay of elements that formed inside a star.

As a pre-stellar nebula collapses, any small initial random movement becomes converted into rotation around a central axis. As collapse continues, gravitational and collisional interactions occur between parts of the nebula moving in different directions. If two objects moving in different directions collide and stick, the resulting larger object moves in a direction that is the average of the two original directions. If the bodies are different sizes, the post-collision direction is closer to the original path of the larger body. This results in flattening the nebula from a roughly spherical shape into a disk as motion differing from the average motion of the nebula is quickly damped out. In addition, the presence of gas causes a drag on the dust present in the nebula, which causes it to drift inward toward the nebular center. At the center of the disk is a rapidly growing object—the protosun, which will eventually become the Sun. Examples of nebulae in this stage as imaged by the Hubble Space Telescope can be seen in Figure 5.1, and a schematic diagram of the process is shown in Figure 5.2. While the story of how the Sun (and stars in general) formed is a fascinating one, we must now turn our attention from this central attraction to the small amount of matter that would not become part of the Sun.

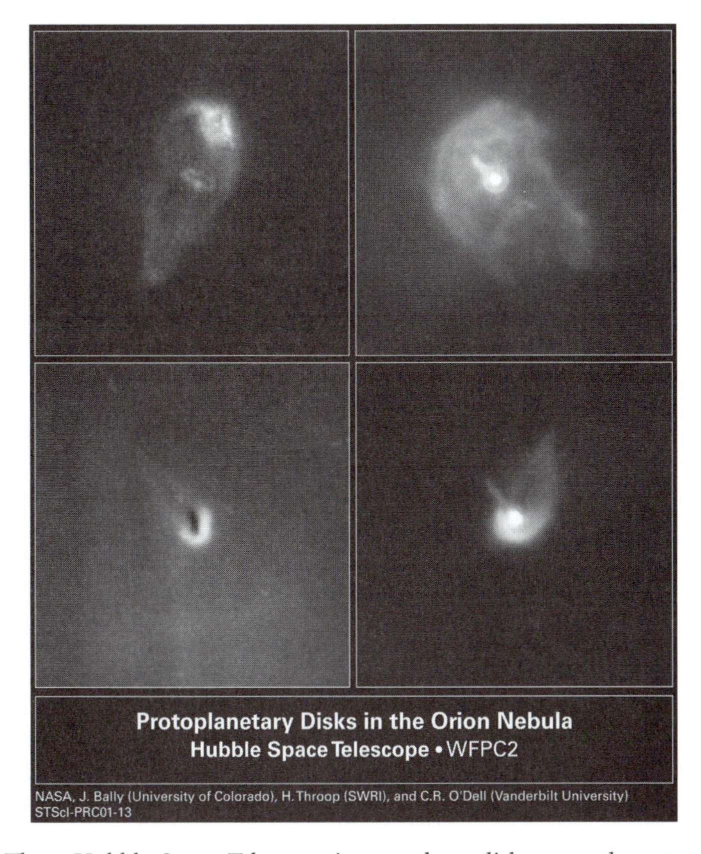

Figure 5.1 These Hubble Space Telescope images show disks around protostars in the Orion Nebula. The still-forming stars at the centers of the disks shine brightly, while the dust and gas is silhouetted against the brighter background. Images like these confirm the general theory of how our solar system formed, and also provide further information to allow scientists to refine and update those theories. NASA.

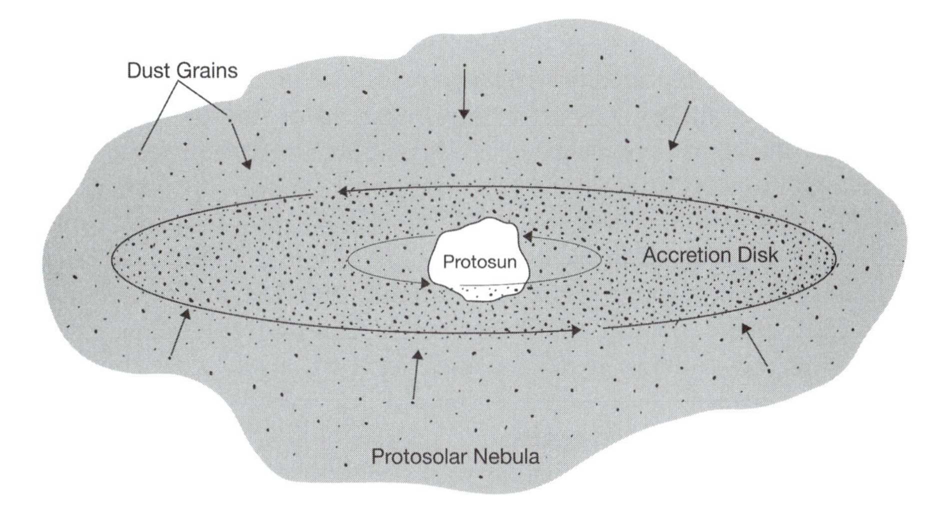

Figure 5.2 This cartoon shows a simplified side view of the protosolar nebula. Dust and gas orbit the forming protosun. Collisions between the dust particles and interactions in the gas causes an originally spheroidal cloud to collapse to a disk. Illustration by Jeff Dixon.

Catastrophic Theories

The nebular theories of solar system origin have only been generally accepted since the latter part of the twentieth century. Competing ideas included the "catastrophic" theories. These theories imagined the solar system as unusual and its formation to be a low-probability event. The most common form of the theory began with a very close pass between the Sun and another star. During the close pass, mass was drawn off of the Sun by the gravity of the passing star, and vice versa. This mass was hypothesized to remain in orbit and eventually coalesce to form the planets. With the advent of computers, astronomers were able to show that any mass drawn from the stars would quickly fall back onto the stars after the close pass. Modifications to these catastrophic theories inevitably were also shown to be inadequate, and with the realization that the nebular hypothesis had no such problems, the catastrophic theories were abandoned.

THE CONDENSATION SEQUENCE

The temperature of the solar nebula varied with distance from the protosun, just as the temperature in the current solar system varies with distance from the Sun. In addition, the nebula as a whole experienced cooling after initially high temperatures. Some compounds and molecules are more stable at high temperatures than low ones, while others are more stable at low temperatures. Experiments have been performed to determine the thermodynamic characteristics of a great many minerals, and given a starting composition like the Sun, and a temperature and pressure (which varied at different times and places in the solar nebula), geochemists can calculate which minerals will form solids and which compounds will remain as gas.

The materials that form solids at the highest temperatures (1500 K or higher) are called **refractory**. These materials, including aluminum oxide and calcium titanium oxide, were the first to form as the nebula cooled, and could potentially form everywhere in the nebula. As a result, refractory minerals could be expected to be present on or in all of the objects in the solar system, save those where later events changed or destroyed them. The calcium-aluminum-rich inclusions (CAIs) found in carbonaceous chondrites and discussed in Chapter 4 are examples of refractory minerals, and are the oldest solid material known in the solar system. As cooling continued, more solids could form, first metals like iron and nickel, followed by magnesium-rich olivine and pyroxene around 1300 K. Again, these materials could potentially form anywhere the temperature would allow it. This set of minerals—olivine, pyroxene, and metal—is largely what chondritic meteorites are made of, consistent with other evidence that these are among the oldest rocks in the solar system.

It is thought that the region of the solar nebula where the terrestrial planets formed was dominated by material with roughly chondritic composition. The chondrites are thus quite literally the material from which the terrestrial planets were built. This would suggest that Venus, Earth, and

Mars should have roughly the same composition when considering all of their mass: core, mantle, and crust. By looking at the elements and minerals present on the Earth's surface and the few mantle rocks that have been found, and comparing those to the chondrite meteorites, geochemists can calculate the most likely composition of the deep interior of the Earth.

Hydrogen is not found in refractory materials. The large central mass that would become the Sun had enough gravity to hold hydrogen gas in place, but the dust grains forming in the inner solar system could not keep their hydrogen. At a certain distance from the Sun, however, the temperatures were cool enough to allow hydrogen-bearing compounds to form. First, hydrated minerals form at around 500 K. These minerals contain water or hydroxyl (OH) as part of their chemical structure, with some having 10 percent or more water by weight. Hydrated minerals are commonly found on the Earth today, including such well-known examples as talc, gypsum, and asbestos.

Further still from the Sun, where the temperature reaches roughly 200–250 K, water ice can form, as depicted in Figure 5.3. Once we reach that distance, sometimes called the **frost line** or **snow line**, the amount of mass that can condense from the nebula goes up immensely. Hydrated mineral formation is limited by the abundance of magnesium and silicon, neither of which is nearly as abundant as hydrogen and oxygen. When all of the magnesium and silicon has been incorporated into mineral grains, no further hydrated minerals can be made regardless of how much hydrogen or oxygen is still free. Water ice formation, on the other hand, is limited by the much greater amount of oxygen, so much more can be formed. The exact solar

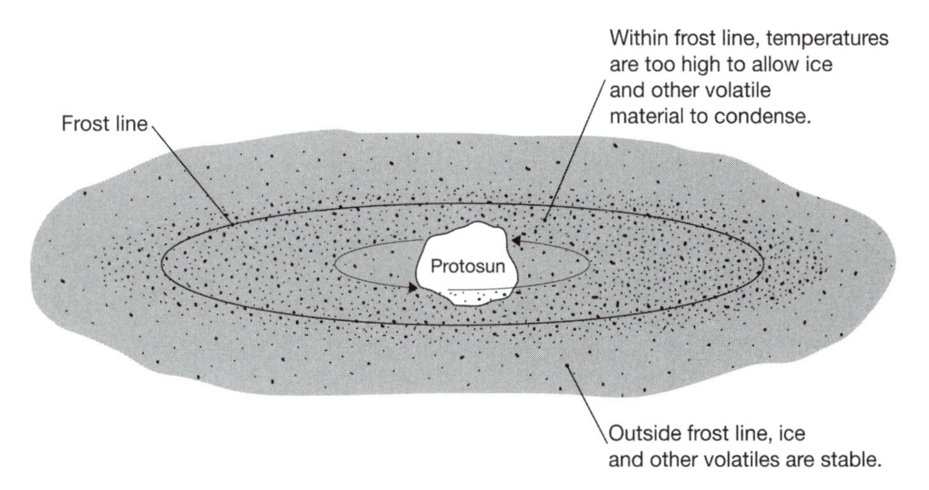

Figure 5.3 Because of differing temperatures, different compounds were stable depending on location in the solar nebula. Closer to the forming sun, only silicates and metals can condense and be incorporated into the forming planetesimals. Beyond the distance where water ice becomes stable, icy objects can form. This "condensation sequence" accounts for the different compositions of different planets and the different natures of the asteroids and comets. Illustration by Jeff Dixon.

distance of the frost line is still under debate, but is thought to be near 5 astronomical units (AU), or five times the distance from the Earth to the Sun. Unsurprisingly, this is where Jupiter, the largest planet in the solar system is found.

Beyond the frost line, additional important compounds condense as well. We expect **volatile** materials, or those solids with low boiling temperatures like carbon dioxide and methane, to have formed in the outer solar system at low temperatures. In an idealized case, therefore, we would expect an object in the far reaches of the solar system like Eris or Sedna to be composed of a mixture of all of the condensable materials in the solar system: refractory solids, chondritic silicates, water ice, and other volatile solids. Large objects like Neptune would have these materials as well as the gravity to be able to hold on to large amounts of hydrogen and helium gas.

This sequence of material creation at differing temperatures in the early solar nebula is called the **condensation sequence**, with the process called **equilibrium condensation**. The actual compositions of objects in the solar system differ somewhat from what would be predicted by the condensation sequence due to mixing, heterogeneity in the nebula, and random events associated with further evolution. However, it does a good job of explaining roughly what should be found in the solar system as a function of solar distance, shown in Figure 5.4. It can also be used to estimate the compositions expected for extrasolar planets based on their distances from their central stars, making some assumptions about the initial mix of gas and dust.

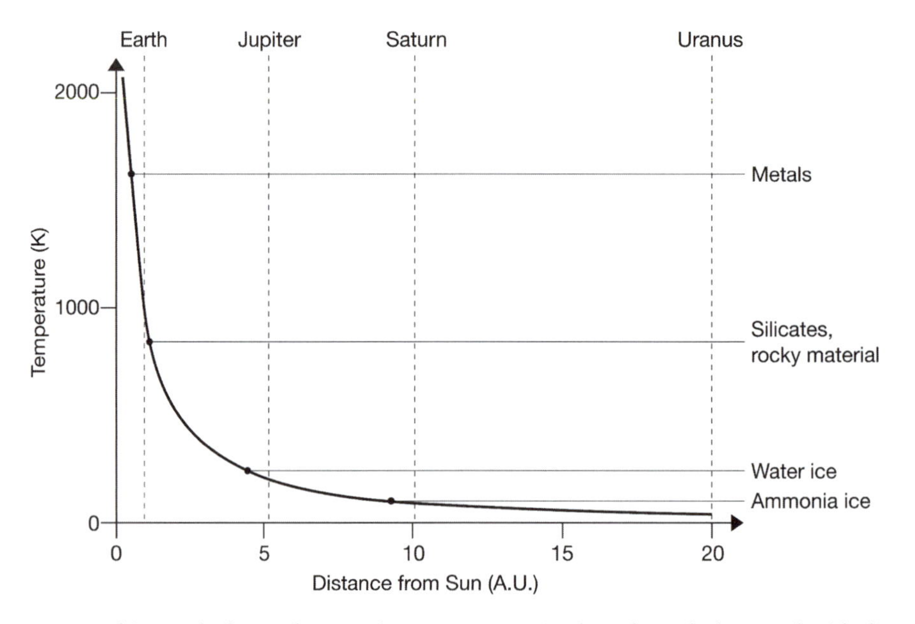

Figure 5.4 This graph shows the way the temperature in the solar nebula varied with distance from the Sun. As shown more qualitatively in the previous figure, different materials are stable at different distances. At the distance of the asteroid belt (about 2–4 AU, only metals and silicates are stable. At larger distances, where comets were formed, water ice, ammonia ice, and other compounds are also stable. Illustration by Jeff Dixon.

ACCRETION AND PLANETESIMALS

The minerals condensing from the solar nebula were very small, only roughly the size of dust. They were also orbiting the Sun in the presence of gas, which slowed them down by gas drag and sent them spiraling slowly inward, as mentioned before. When dust encountered other dust, it was moving slowly enough that collisions were gentle and the pieces tended to stick to one another, or accrete. The angular, jagged shapes of condensing dust also helped stick pieces to one another. As accretions of dust got larger, they would have a greater cross-sectional area and a greater opportunity for further growth. As they grew larger, the gas drag would grow less effective—the force due to gas drag increases with the area of a body (which is proportional to the radius squared), but the mass of the body increases more quickly (proportionally to the radius cubed), so the acceleration (equal to the force divided by the mass) gets smaller. As larger accretions became less affected by gas drag, they had even more opportunities for accreting smaller accretions that were still spiraling in due to gas drag.

Eventually, accretions reached the size at which their gravity started to affect the other bodies around them. This happens at a diameter of roughly 1 kilometer. At this size, they are conventionally called **planetesimals** by astronomers. The first planetesimals to form had a great advantage over smaller accretions in terms of capability for growth, and as they continued to grow this advantage increased. Because of this, the process by which planetesimal mass and size increased is called **runaway growth**. At these sizes, gas drag can no longer change the orbits of planetesimals. Smaller objects continued to spiral in toward the Sun, however, providing a steady supply of objects for the planetesimals to accrete and allowing them to grow larger still. Within a few million years, some planetesimals grew to be **planetary embryos** hundreds of kilometers in size and would go on to form the major planets.

Accretion rates were potentially high in the inner solar system, but as discussed already, the temperatures were sufficiently high that the amount of accretable solids was limited. In the outermost solar system, temperatures were low enough to allow a wide range of solids, but the accretion rates were extremely slow, discouraging the creation of large planets. At a great enough distance, the density of the solar nebula eventually reached a value where accretion stalled out because the time between collisions was simply too long. This distance is thought to have been reached beyond Neptune, where the current transneptunian object (TNO) population represents the original planetesimals that never formed into planets. Some objects did manage to grow to planetary embryo size before stalling out, notably Pluto and Eris.

As noted before, Jupiter sits at a "sweet spot" where silicates and water ice are solids, but also where the accretion rate was relatively high. While these were ideal conditions for growing a large planet, they were also optimally bad conditions for Jupiter's would-be neighbors. Proto-Jupiter's rapidly growing gravitational pull influenced the orbits of the planetesimals nearby, increasing the eccentricity of the orbits, which increased the average collisional speeds between those objects. Proto-Jupiter itself was large enough to remain intact and continue accreting even with higher encounter speeds, but the smaller objects in what is now the asteroid belt were not massive enough to have sufficient gravity to pull themselves back together after a large collision, and encounters on average became corrosive rather than accretionary. As in the Kuiper belt region, some objects (like Ceres, Pallas, and Vesta) did manage to make it to planetary embryo-size before Jupiter's growth preempted their own. And again, as in the Kuiper belt region, the majority of the population represents the original planetesimal population, though in this case with a composition more representative of the inner solar system rather than the outer solar system.

The composition of the major planets is an average of the compositions of the objects that accreted to form them. While measuring the compositions of the deep interiors of the major planets can only be done indirectly at best, the fact that the asteroids and comets are the building blocks of the large planets means that they can be used to tell us about the compositions of those larger bodies. This concept is discussed further in Chapter 4.

Heating and Melting in the Solar Nebula

A very early heating event is required to explain the creation of chondrules, the millimeter-sized bits of melted silicate that give chondrites their names, as discussed in Chapter 4. However, there is ample evidence in the meteorites for additional processes driven by heat. Many chondrites show evidence of metamorphism, the reactions between water and rock that we see evidence of in carbonaceous chondrites required some heat to begin, and achondrites all experienced temperatures hot enough to melt rock. Where did this heat come from?

Today, the rocky and icy bodies of the solar system generate heat through the decay of radioactive elements. After more than 4 billion years since the solar system began, only a few radioactive elements (most notably uranium and thorium) have half-lives long enough to remain in abundances great enough to generate much heat. Early in solar system history, however, short-lived radioactive elements could still be found.

Evidence from meteorites shows the supernova (or supernovae, if more than one) that began the nebular collapse through shock waves also contributed short-lived radioactive elements. These elements, like ^{26}Al and ^{60}Fe, have half-lives short enough that they rapidly disappeared, but they gave off large amounts of heat as they did so. While we cannot find ^{26}Al or ^{60}Fe in meteorites today, researchers have found the atoms they decayed into or "daughter products," supporting the idea that they were responsible for the heating.

THE END OF ACCRETION

Roughly 10 million years after the Sun began fusing hydrogen, it is thought to have emitted a strong solar wind, which removed the remaining gas and much of the remaining dust from the solar system. This is called the **T-Tauri stage**, after the first star observed to exhibit this wind. As smaller objects stopped experiencing gas drag, this had the effect of significantly slowing accretion.

As the planets reached their final sizes, there were still a number of planetary embryos (as well as planetesimals) present on planet-crossing orbits. As discussed in Chapter 3, objects in these orbits do not last long before impacting other objects or being ejected from the solar system. It is thought that over 90 percent of the original mass in the asteroid belt was removed by Jupiter through close encounters, though the exact timing of this removal is a matter of some debate. Planetesimals in the outer solar system were also removed by the giant planets. In their case, however, current models suggest they remain bound to the Sun and form the population in the Oort cloud, the region stretching nearly halfway to the nearest star where long-period comets are thought to originate. Surprisingly, then, these furthest members of our solar system are thought to have originated significantly closer to the Sun than the Kuiper belt.

The giant planets had a great influence on the orbits of the remaining small bodies at the end of accretion. Conversely, while a single encounter with a planetesimal or planetary embryo had no measurable effect on giant planet orbits, the sheer number of small bodies undergoing encounters had a large cumulative effect. Jupiter, on average, scattered objects out of the solar system to larger orbits, which had the effect of moving Jupiter itself to a smaller orbit. Saturn, Uranus, and Neptune, on the other hand, scattered objects into smaller orbits on average (where they tended to encounter Jupiter) so that those objects slowly moved outward. As Neptune moved (perhaps as much as several AU), it "captured" some planetesimals into resonances. This is thought to be the origin of the populations of the Neptune Trojans (objects orbiting 60 degrees leading and trailing a planet) and plutinos (objects with orbits like Pluto's, in a 3:2 resonance with Neptune), and perhaps the Jupiter Trojans as well.

It has also been suggested by modelers that the migrating planets themselves could have entered resonances with one another. As Saturn and Jupiter move, for a time their orbits will be in a 2:1 resonance. This would have a massively destabilizing effect on orbits in the rest of the solar system, and would have resulted in a rapid depletion of the small body population from the amounts expected early in solar system history to the amounts seen today. The timing of this event also seems to coincide with the **terminal lunar cataclysm**, an upsurge in the amount of basin formation on the Moon (and presumably the Earth, though the evidence here has been erased after billions of years). While neither the extent of the terminal lunar cataclysm nor the exact dynamical models of migrating planets are universally accepted, further work will test these hypotheses.

THE AGE OF THE SOLAR SYSTEM

The earliest scientific estimates of the age of the Earth, and of the solar system, were limited by the experiences available to scientists. Geologists studying the creation of sedimentary rocks recognized that at least hundreds of millions of years or longer were necessary given the pace of current sediment transport. Biologists also considered that similar amounts of time were required for Darwinian evolution to create the diversity in species seen today. Physicists of the nineteenth century, however, knew nothing about nuclear fusion and theorized that the Sun was powered by contraction, converting potential energy into heat. They calculated that the Sun could not possibly live for longer than 20 million years, given that heat source. Similarly, before the discovery of radioactivity, it was thought that the increase in heat with depth below the Earth's surface was due to energy from the planet's formation, which again implied a short time since creation.

With the discovery of nuclear fusion, it was realized that the Sun could be billions of years old. And the earlier discovery of radioactivity and its measurement in rocks led scientists to understand that the Earth itself was billions of years old. These same techniques when applied to meteorites showed that these objects are older than the oldest Earth rocks yet found—4.5 billion years old. Even those objects that have experienced melting have been solid for practically the entirety of solar system history. In combination with the study of the textures of meteorites, further discussed in Chapter 4, meteorites have been recognized as the oldest rocks in the solar system, and the CAIs they can contain are seen as the oldest solid matter. Often, other materials are dated using the CAIs as an initial reference time.

THE SMALL BODIES AND SOLAR SYSTEM FORMATION

The asteroids, comets, and dwarf planets are all of central importance for researchers studying the formation of the solar system and its earliest history. The most obvious reason is because they are leftover planetesimals (or embryos in the case of Ceres, Pluto, Eris, and a few others). But there are other reasons as well.

Their orbits give insight into the positions of the major planets. Simulations and calculations that try to reproduce the current orbits of small bodies have supported the idea that the giant planets were migrating in early times. For instance, the existence of the Trojan asteroids of Jupiter and Neptune are much easier to explain if those planets changed their semimajor axes early in solar system history. The presence of asteroid collisional families shows that the asteroid belt was more densely populated at one time, since the likelihood of large impacts today is very small, and the probability of creating the many large families we see is vanishingly small. With the information that there once was more mass in the asteroid belt, models of solar system formation can be made more accurate. Furthermore, that information naturally leads to the question of where that mass went, again leading to the idea that the major planets ejected the mass and have migrated as a result.

Second, the major planets are so large and have experienced so much processing and mixing that clues from their formation can only be indirectly found after painstaking work, if at all. The fact that the small solar system bodies are much closer to pristine objects gives us the opportunity to study the creation of the solar system, and the Earth specifically. The samples of small bodies in our collections provide scientists with material that has been almost unchanged for billions of years. Telescopic observations can provide compositions for asteroids and comets, discussed further in Chapter 7, which can then be used in conjunction with the meteorites to map the changing composition of the protoplanetary nebula with solar distance. However, while significant insights into solar system formation have been gained over the past decades, both meteorite evidence and telescopic observations have limitations—most meteorites have spent significant time on the Earth, where they have been changed by weather and erosion in ways that are sometimes difficult to unravel. Compositional estimates using telescopic data can suffer from uncertainties in calibration and sometimes poor data. For these reasons, many scientists support space missions to collect fresh material from asteroidal or cometary surfaces and return it to Earth where it can be studied. At least one such mission, discussed further in Chapter 12, is likely to be completed in the next decade. However, given the great diversity of material in the solar system and the cost of space missions, it seems certain that the vast bulk of our knowledge will continue to be gained via meteorite study and remote sensing.

SUMMARY

Through observations of star-forming regions and what appear to be planetary systems around other stars in their early stages, and via theoretical models, astronomers have a relatively robust idea of how our solar system formed. According to the models, a large cloud of gas and dust collapsed into a disk, with the protosun in the center. The solids in the disk collected into ever-larger objects, with the largest objects growing more quickly than the smaller ones. While the largest objects eventually grew large enough to become the major planets, some bodies stalled out at much smaller sizes. Today's asteroids, comets, transneptunian objects (TNOs), and dwarf planets are mostly the remnants of that material, which was never collected into planets and retains the physical properties it had when the solar system formed. Because of the great changes that have occurred on the major planets since then, the small bodies of the solar system are the objects that best represent the materials and conditions present in the earliest times throughout the solar system.

RECOMMENDED READING

University of Arizona Space Science Series: This series of books covers the formation of the solar system in great detail. The most recent volume is *Protostars and Planets V*, published in 2007.

WEB SITES

Web site for the Hubble Space Telescope. A great deal of data from the Hubble Space Telescope has been used to study protoplanetary disks around other stars. The site contains many of these images as well as explanations for them: http://hubblesite.org.

The "Windows to the Universe" Web site includes a set of pages about solar system formation at various levels of detail (in Spanish as well as English), and a reading list that includes books on solar system formation: http://www.windows.ucar.edu.

6

Sizes, Shapes, and Companions of Small Bodies

In this chapter, we ask some of the most fundamental questions about the small bodies of the solar system: How big are they? What shapes do they have? Do they have moons? We will discuss the answers to these questions, as well as the techniques used by astronomers and planetary scientists to find those answers.

SIZING AND SHAPING UP SMALL BODIES

In our everyday lives, we measure the sizes of objects constantly. Sometimes it is done directly with a ruler, but often we do it subconsciously, when we look up at a tall building for instance, or note the height of those we pass on the street. The sizes of asteroids, comets, and dwarf planets are of great interest to scientists for a number of reasons. Knowing the size of an object gives a measure of the strength of its gravity and also insight into its internal structure, the temperature its interior is likely to have reached, how dangerous it would be if it were to strike the Earth, what compounds are likely to be stable on its surface, and numerous other models that are critically dependent upon size. Indeed, the definition of a planet is indirectly based in part on the size of the object.

How do we measure the size of small bodies? The most obvious tactic would be to take a picture or image of the object of interest and measure its angular size. When combined with the distance to the object, the diameter in kilometers can be calculated. This technique was used to determine the sizes of the planets as early as the late-eighteenth century, using a device called a micrometer to measure an object's angular size in the days before photography or digital cameras. Other things being equal, a larger telescope enables measurements of smaller and smaller objects. The unaided human eye has a theoretical resolution of roughly 0.5–1 arcminute (an arcminute is 1/60 of a degree), or about 0.01 degree. The Hubble Space Telescope's theoretical resolution is 5,000 times greater than that.

The presence of the atmosphere creates a limit to the angular size that can be measured from the ground, regardless of telescope size. Instabilities in the atmosphere cause the appearance of objects to shimmer or twinkle, an effect with the ungrammatical sounding name of **seeing**. Seeing limits measurements to a few arcseconds for much of the Earth, though at astronomical observatories specially chosen for good seeing, this number can go as low as 0.25–0.5 arcseconds. For objects with a large angular extent like the Moon, or the planets, or some galaxies or nebulae, the effect of seeing is relatively minor. For objects of very small angular extent, like stars, planetary satellites, and small bodies, the seeing dominates measurements from the ground.

The major planets in the solar system are large enough that their sizes can be easily measured directly, and by the 1800s their diameters were known with only relatively small uncertainties. When measurements of asteroids and comets were attempted, however, problems with the direct measurement approach were found immediately. For comets, the diameter of the coma is not fixed, but varies depending on temperature and distance to the Sun, so it is not a useful quantity for measurement. In addition, the coma does not have a sharp edge, making its extent unclear. The diameter of the nucleus is a much more interesting quantity, but is difficult to measure with the coma interfering. These factors, combined with the small size of the nucleus, have frustrated measurements of cometary sizes until relatively recently.

Asteroids are by and large too small to measure directly. Ceres, the largest object in the asteroid belt, only appears as large as a golf ball would at a distance of 1.5 km, with the next-largest objects, Pallas and Vesta, only roughly half that size. Pluto appears about 10 times smaller from the Earth than Ceres, with other TNOs appearing smaller still. Two advances have allowed some measurements of the largest of small bodies using the direct technique, however: the Hubble Space Telescope (HST) allows observations to be made without concern for seeing; and advanced telescopes have been developed that use "adaptive optics" to measure and compensate for seeing on very short timescales. The HST images of Ceres and Pluto in Figure 6.1 clearly show their disks. A small number of objects have been visited by spacecraft, which allow sizes to be directly measured relatively easily and very accurately. These include both asteroids and comets, with an image of Comet Borrelly from the *Deep Space 1* mission also shown in Figure 6.1.

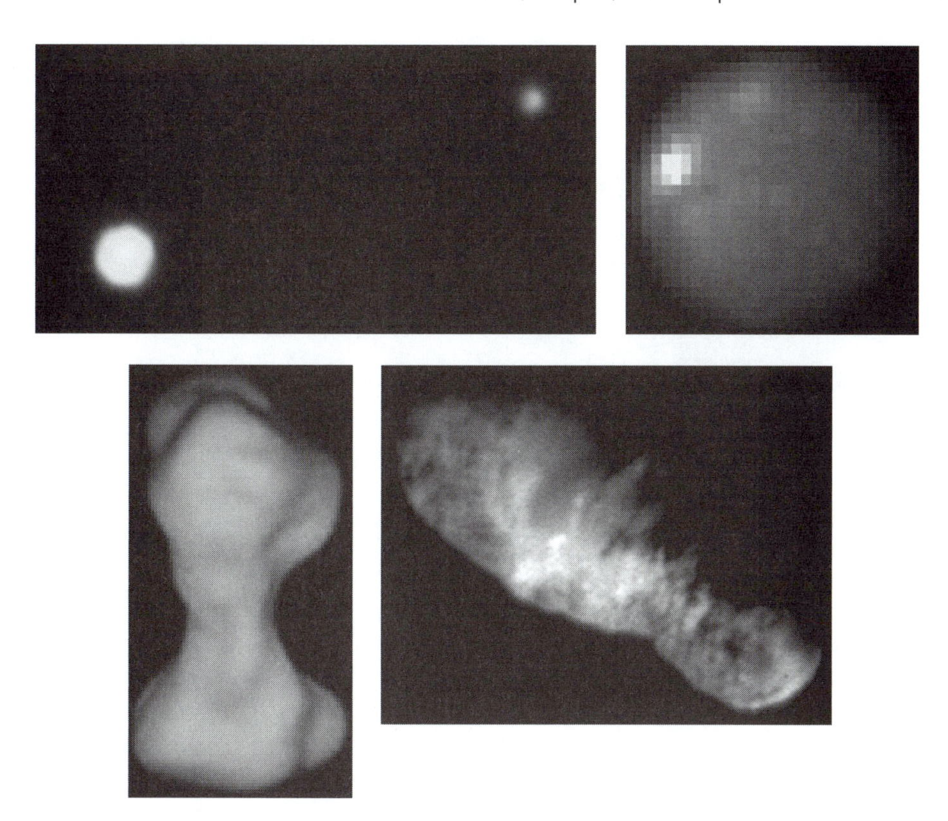

Figure 6.1 These four images show a range of shapes and sizes seen in the small body population. At top left is Pluto, nearly spherical and roughly 1200 km across, and its satellite Charon, about half its size. At top right is Ceres, midway in size between Pluto and Charon, and also nearly spherical. The Pluto, Charon, and Ceres images were all taken using the Hubble Space Telescope. At bottom left is a model of the asteroid Kleopatra, made using radar data. Kleopatra is 217 km long, with the shorter axes 94 and 81 km. At bottom right is Comet 19/P Borrelly, with the image taken by the Deep Space 1 spacecraft. The comet's visible area is 8 by 3 km. Kleopatra and Borrelly are distinctly irregular in shape, Kleopatra's shape in particular comparable to a dog bone. Top left, Dr. R. Albrecht, ESA/ESO Space Telescope European Coordinating Facility; top right, NASA; bottom left, NASA/JPL/Northwestern University; bottom right, NASA/JPL.

Objects that are large enough to show their shape are called **resolved**. The majority of asteroids and TNOs are **unresolved** regardless of the telescope that is used, appearing only as points of light. Even in this case, their sizes can be measured or estimated by using other techniques.

The technique that may be most familiar is simply measuring the object's brightness, a technique called photometry. Looking out over a valley at night, we can roughly judge the relative distance between two houses by noting how bright their lights are. Similarly, we can estimate the changing distance to a motorcycle at night by watching its headlight change in brightness. The brightness of a small body depends upon three factors: its distance from the Sun and Earth, its size, and the darkness or lightness of the minerals on its surface. This last factor is called albedo. There are several different

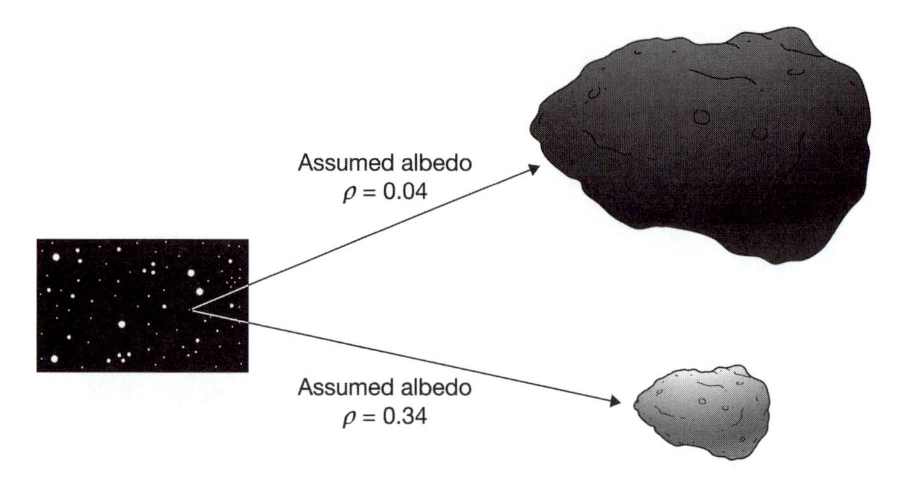

Figure 6.2 For a given brightness, an unresolved small body can either be very reflective and small (like the bottom object) or larger and darker (like the top object). Without information about the albedo (or reflectivity) of the object, astronomers are required to make an estimate of the albedo and take note of their uncertainty. If an albedo measurement can be made, the true size of the object can be determined. Illustration by Jeff Dixon.

kinds of albedo, which measure slightly different things, but all represent the fraction of light falling on a surface that gets reflected; an object with an albedo of 0.30 reflects 30 percent of the light that falls on it. The brightness, or flux, of a body increases with albedo, and with its area (and thus its radius or diameter). A small, high-albedo object can be just as bright as a large, low-albedo object at the same distance, if both are unresolved. This is schematically shown in Figure 6.2. The following equation quantifies this relationship:

$$\text{Flux} = S \times \text{albedo} \times \text{area}/(a^2 \times \Delta^2) = S \times \text{albedo} \times \pi r^2/(a^2 \times \Delta^2)$$

where S is a constant related to the brightness of the Sun, a is the distance between the object and the Sun, and Δ is the distance between the object and the observer.

As will be discussed further in this section, albedos for small bodies can vary greatly. Unfortunately, they are more difficult to measure than brightness. Astronomers commonly treat objects without known albedos in one of two ways. They can assume or guess at the albedo based on other factors, and use that value to estimate a size, or they can simply report the brightness of the light reflected from the object, which is measured in **magnitudes**. Because brightness changes with distance from the Sun and Earth, and because those distances for an object are readily calculated if its orbit is known, any measured magnitude can be extrapolated to a particular set of distances to allow easy comparison among many bodies, resulting in an **absolute magnitude (H).** The absolute magnitude represents how bright a small body would be at 1 AU from the Earth and 1 AU from the Sun. For historical reasons, smaller magnitudes represent brighter objects. While two objects with similar absolute magnitudes may have very different sizes depending on their albedos, a large sample of bodies with similar absolute magnitudes will on average have about the same diameters.

Because small bodies move with respect to the stars, there are occasions when they move in front of stars and block their light. These events, called **occultations**, provide another way of measuring the size of an object. Knowing the speed of a body in its orbit (again, readily calculated for objects with known orbits), the duration of an occultation provides a minimum size. Observations of the same occultation from multiple sites can provide additional data on the body's size. Unfortunately, occultations are relatively difficult to observe, since any particular event can only be seen from a small area (much like solar eclipses are only visible from certain places), and the best location for observing might not be known until a short time before the event. Occultations of larger asteroids, Pluto, and Charon have been observed in the past few decades. No comet or TNO occultations (other than Pluto) have been observed, though opportunities for TNO occultations in particular are eagerly sought by astronomers.

Radar observations are particularly powerful for characterizing the near-Earth object population. The receiving dishes used in radar can be built to be much larger than optical or infrared telescopes. In addition, the atmosphere interferes very little with radar. The main limitation of radar is the amount of power required to transmit the signal. The signal loses strength both on its way to the asteroid and on its return to the Earth. A doubling of distance to an NEO requires a factor of 16 increase in power to achieve the same quality of data.

Radar images are different from normal images. The radar signal is sent from Earth at a known time with a known frequency. That frequency has a Doppler shift upon reflection off of the object due to the object's rotation. The time it takes the signal to arrive on Earth allows the distance to the body (or its range) to be very accurately measured. The change in frequency is also accurately measured, and provides the speed with which the body is rotating. Different points on an object's surface are moving at different speeds (for instance, one edge is moving away from the Earth, and the other toward the Earth), and are also at slightly different distances. Therefore, sophisticated modeling can be required to analyze radar data.

Astronomers have also developed techniques to measure small body albedos. As noted before, albedo measurements in conjunction with the known distance to an object and its measured brightness allow a size to be determined. The primary way albedos are measured is by determining the temperature of an object. Imagine two objects, identical except that one is white, and the other black. The same amount of sunlight falls on both. The white object reflects the light that falls on it; the black one absorbs it. As a result, the black object gets hotter and remains hotter as long as they are in sunlight. This effect can be recognized by those who like to walk barefoot—lighter-colored sidewalks are cooler than asphalt-covered parking lots. The warmer an object is, the more infrared light it emits.

The primary technique for measuring albedos is by observing objects in the infrared spectral region, at wavelengths of 10–20 μm (for comparison, the human eye is sensitive to light from roughly 0.4–0.7 μm). At typical temperatures for asteroids, comets, and TNOs, most of the light observed is emitted black-body radiation rather than reflected solar radiation. How does this allow albedos to be measured? By measuring the brightness of objects in the 10–20 μm region, their temperature can be calculated, and from the temperature and their distance from the Sun, their albedo can be calculated. As described previously, knowing the brightness at visible wavelengths, distance to the object, and albedo are all that is required to calculate a size.

Why are sizes useful? As mentioned already, for individual objects the size is an important number to include in models for their history, whether objects have melted or not, how long they are expected to last before having a disruptive collision, and how much damage they might do if they hit the Earth. Knowing the proportion of larger to smaller asteroids and comets is also useful for testing theories and models. This proportion is usually represented as a **size-frequency distribution (SFD)**. As time goes on, objects undergo collisions that can break them into a few pieces, or even disrupt them into countless tiny remnants. The most destructive collisions occur between objects of roughly the same size. Because larger objects are rarer than small ones in the asteroid and TNO populations, a given object will experience many more collisions with smaller objects compared to larger ones. The larger objects have very little danger of encountering an object large enough to disrupt them, since few of those objects exist. In addition, the larger objects have more gravity, which helps them to better resist disruption if an impact does occur. These facts can be used to study the small body population using the current SFD and determine what the original SFD must have been 4.5 billion years ago. The SFDs for both main-belt asteroids and TNOs have been shown to be what is expected for a population that has been experiencing mutual collisions for billions of years.

The Smallest of Small Bodies

The largest of the dwarf planets have diameters of thousands of kilometers. The largest of NEOs have diameters that are upwards of 10 km. By extrapolating the SFD of the small body populations, we expect millions of objects smaller than that to exist. While most of those tiny bodies are too faint to be seen, some have been spotted while making close passes to the Earth. The smallest individual object to be spotted in orbit around the Sun is 2006 QM111, which passed the Earth at less than half the distance to the Moon. Based on its brightness, 2006 QM111 is thought to be 2–5 meters in size, which is not much larger than a human being and would easily fit in most classrooms. Much smaller objects commonly strike the Earth but were not spotted before entry until late 2008, when 2008 TC3 was discovered a day before its impact.

As can be seen from Figure 6.1, asteroids and comets come in a large variety of shapes, from almost round to highly irregular. As with sizes, the most straightforward way to measure small body shapes is by direct imaging. And, again, very few objects have had their shapes measured in this way. The radar imaging described here provides an excellent way to determine the shapes of NEOs. Roughly 30 near-Earth objects have had their shapes measured using radar.

Most shapes that are available were modeled using photometric data. As discussed before, the brightness of an object is dependent upon its cross-sectional area. A perfectly spherical body will always have the same brightness as it rotates. Objects with different shapes, however, will change their brightness as they rotate. The change in an object's brightness with rotation is called its lightcurve. Lightcurve observations at one point in a body's orbit can give the relative sizes of the axes of the cross-sectional area. Observations over several years allows different parts of an object to be seen. When all put together, the shape of an object can be modeled.

Shapes are important for what they tell us about the interior structure of a body. For instance, knowledge of the shape of Ceres from HST observations has allowed astronomers to deduce that body has an internal ice ocean. The shape of an object is also a critical component for classifying it as a dwarf planet or small solar system body. The utility of knowing the shape of small bodies is explored in further detail in Chapter 9.

SATELLITES

Like their larger planetary cousins, some asteroids have been found to have satellites. Through the last decades of the twentieth century, there had been great controversy among planetary scientists as to whether or not moons around asteroids should be commonplace, rare, or even absent. After hints from lightcurves suggested the presence of asteroid satellites, the first confirmed asteroidal satellite was found in 1993 by the *Galileo* spacecraft during its flyby of Ida. This satellite, named Dactyl, led to a flurry of observations dedicated to searching for more satellites and research designed to study the origins of these multiple systems. Most systems have a single large body, called the **primary**, with the satellite or satellites called **secondaries**. In the case of Ida's satellite, Ida is the primary and Dactyl is a secondary.

Some transneptunian objects are also known to have satellites. The most famous of these systems is the Pluto-Charon system. Charon was discovered in 1978 and is nearly the same size as Pluto, leading some to consider the pair a "double planet." In 2005, two more satellites were found around Pluto by researchers using images from the Hubble Space Telescope, shown in Figure 6.3. The dwarf planet Eris is also known to have a satellite, named Dysnomia, leaving Ceres the only dwarf planet not known to have a moon. In total, over 100 small bodies with satellites have been discovered,

Figure 6.3 Hubble Space Telescope images of the Pluto-Charon system discovered two additional moons of Pluto in the year 2005. Multiple systems like this system are rare. NASA/ESA/H. Weaver (JHUAPL)/A. Stern (SwRI)/HST Pluto Companion Search Team.

including dozens of NEOs, main-belt asteroids, and TNOs. In addition to Pluto, at least two asteroids are known to have multiple satellites.

While direct imaging can only rarely provide a measurable size or shape for a small body, it has been an effective technique for discovering satellites. Often, an occulting disk is used to block out the light of the central body, which can be hundreds of times brighter than any satellites. With the central body's light blocked, any faint objects that may be present can be spotted much more easily, and tracked through an orbit, as with the satellite of 45 Eugenia seen in Figure 6.4.

The same kinds of lightcurve observations that can be used to determine the shape of an object are also used to determine if satellites are present. As a primary and secondary orbit the Sun, there are times when one blocks the other along the line of sight to the Earth. These so-called **mutual events** are seen in the lightcurve as relatively sudden and sharp drops in the brightness of the system. There are some configurations that are difficult to definitively identify as satellites as opposed to shapes—a pair of similar-sized, very closely orbiting objects has a lightcurve that is close to that of a dumbbell-shaped single object, for instance. Due to the relative ease of collecting lightcurves, however, this technique has allowed many candidate binary systems to be identified.

Observations using radar have been particularly useful in finding companions for small bodies. Because a satellite moves in a very different way relative to the Earth than its primary, satellites are easily detectable in the range-Doppler data. One of the most remarkable radar

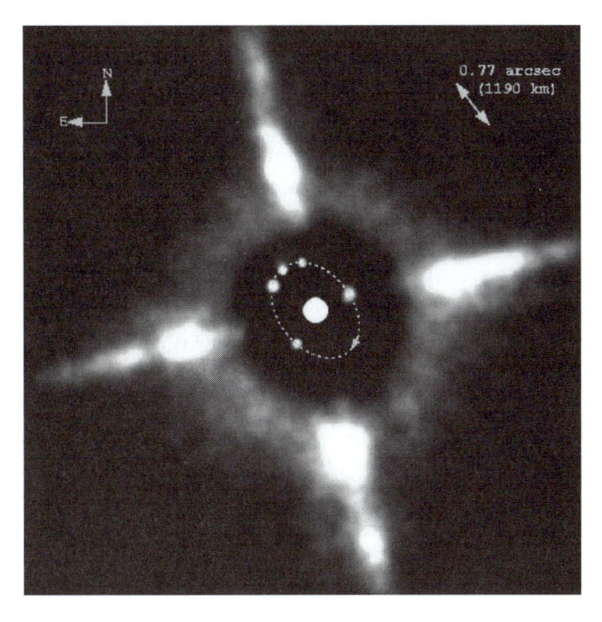

Figure 6.4 This image shows the orbit of the satellite of Eugenia, named Petit-Prince. Eugenia itself is at the center. Images were taken at several different times, each showing the satellite in a slightly different place. All of those images were superimposed to create the image above. Also superimposed is the orbit of Petit-Prince as a dashed line, and an arrow showing its direction of motion. The cross-shaped pattern is due to stray light in the telescope. AP Photo/European Southern Observatory (Laird Close).

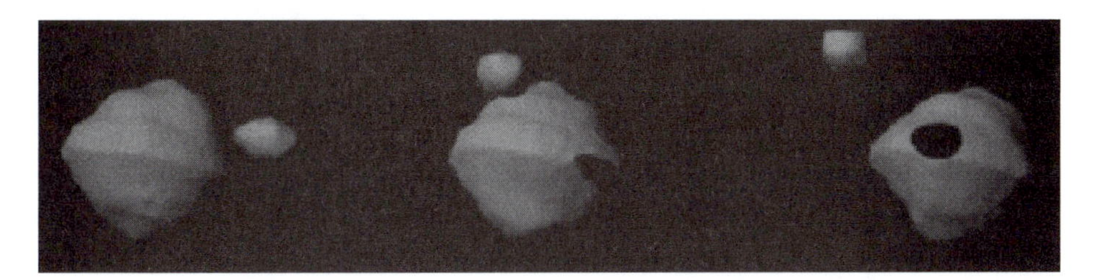

Figure 6.5 Three different views of the same satellite system are shown here. The primary is 1999 KW4, a near-Earth asteroid with a diameter of 1.2 km. The figure depicts a view modeled using radar data, rather than being an actual image. This asteroid is rotating very quickly, near the speed at which its own gravity could not keep it together. Most binary NEOs are very fast rotators, which has led to theories that gravitational interactions with planets or thermal forces "spin up" NEOs until they lose mass and satellites form. Also notable for 1999 KW4 is the apparent ridge near its equator. Calculations show that the rapid spin and gravitational pull of the satellite both tend to move material toward the equator, where it accumulates. AP Photo/NASA/JPL.

observations of a satellite system is the object 1999 KW4, shown in Figure 6.5. This object is rotating quite rapidly, with a period of less than 2.5 hours. The rapid rotation plus the gravitational pull of its satellite steadily moves material toward the equator, where a pronounced ridge can be seen in the figure.

Naming of Satellites

For the major planets, satellite names often are related to the name of the primary. For instance, the largest satellites of Jupiter all are named after paramours of the Roman god. Similarly, small body satellites are often given permanent names that recall the name of the primary. Charon, the first-discovered satellite of Pluto, is named for the ferryman who brought souls into the land of the dead in Greek mythology. The asteroid 243 Ida is named after a mythological nymph who lived on the very real Mt Ida in Crete. Its satellite, Dactyl, was named after other creatures that lived on that mountain in legend. Asteroid 45 Eugenia, named in 1857 after the wife of Napoleon III, has a satellite named Petit-Prince after her son, but also with a nod toward the Little Prince of literature who lived on an asteroid. For objects without permanent names, such as 66391 1999 KW4, satellite names are given additional numbers reflecting the name of the primary and the year of discovery—its satellite is S/2001 (66391) 1. While much less euphonious than "Petit-Prince" or "Charon," it does serve as a unique name.

All these techniques are sensitive to different sorts of satellite systems. Direct imaging techniques are best suited to discovering and observing objects that are well-separated, and work best on brighter (and thus typically larger) satellites. Lightcurve observations are also better suited to larger secondaries (since they affect the shape of the lightcurve more), but can identify close components more easily than direct imaging.

WHAT ARE SMALL BODY SATELLITES LIKE?

The characteristics of small body satellites are very different from those of the satellites of the major planets. They also seem to differ between the near-Earth object, main-belt, and TNO populations. The satellite systems found among the NEOs generally have small, rapidly spinning primaries with primaries having diameters less than roughly 5 km and rotation periods less than 5 hours. The primaries are also usually spherical rather than irregular. The secondaries are most commonly less than about half the diameter of the primary, and their separations tend to be small—only a few times the size of the primary itself (for instance, a 3-km-diameter primary might have a 1-km-diameter satellite orbiting 6 km from the primary's center, though there is variation in all of these ranges). As a whole, roughly 15 percent of NEOs are thought to be binary systems, with the fraction of fast-spinning binary systems much larger (perhaps two-thirds).

By contrast, the percentage of binaries in the main-belt asteroid population is much smaller among surveyed objects. Among the large main-belt objects, only 2 percent are in multiple systems. The size differences between primaries and secondaries in this population are much larger than that of the NEOs, and they are typically at much larger distances. However, the techniques used for studying main-belt asteroids cannot reach the small sizes seen in the NEO population, making direct comparisons difficult. Indeed, if the known NEO binary systems were moved to the distance of the main asteroid belt, they would be very difficult to recognize as binaries. As techniques improve and additional data are collected, astronomers will be able to determine whether the currently known main-belt and NEO binary populations differ because they formed differently or simply because small binary systems are different from large ones.

Binary TNOs are seen to be different from asteroids. Again, care must be taken when comparing systems detected using quite different means, but binary TNOs seem to be largely composed of similar-sized objects in relatively distant orbits from one another. Roughly 5 percent of TNOs are thought to be part of a binary system. Unlike the largest asteroids, few of which have satellites, many of the largest TNOs (Pluto, Eris, 2003 EL61, and 90482 Orcus) have at least one satellite.

Because the satellites of small bodies are difficult to observe, their physical characteristics have been difficult to determine. The small amount of available data for asteroidal satellites suggests that primaries and secondaries are composed of the same material. The most detailed data for a TNO system, that of Pluto and Charon, suggests that Charon has a higher fraction of water ice than does Pluto.

HOW DO SATELLITE SYSTEMS FORM?

By the late-twentieth century, planetary scientists had largely convinced themselves that the small bodies did not have satellites. This was mainly because scenarios developed to explain the formation of satellites for the major planets were not consistent with asteroid-sized bodies. Those theories included miniature versions of the formation of the entire solar system where a planet plays the part of the Sun, invoked to create the large satellites of Jupiter (and not possible around smaller objects); "giant impact" theories where very large collisions threw huge amounts of material into space, later coalescing into a satellite (such impacts would utterly destroy most small bodies); or capture via complex multi-body interactions (exceedingly unlikely given the weak gravity of small bodies and the density of objects in the current solar system).

The discovery of asteroidal satellites has led scientists to revisit their assumptions. While the mechanisms for creating satellites of the major planets are difficult to apply to the small bodies, mechanisms unique to the asteroidal population have been identified. The first is binary creation as a result of impact. Unlike models of the formation of the Moon, which involve the coagulation of molten debris in Earth orbit over a long time, these models consider the gravitational interactions of solid ejecta from an impact into an asteroid. While the vast majority of ejecta mass is either lost entirely or falls back onto the target object, a small amount can remain in

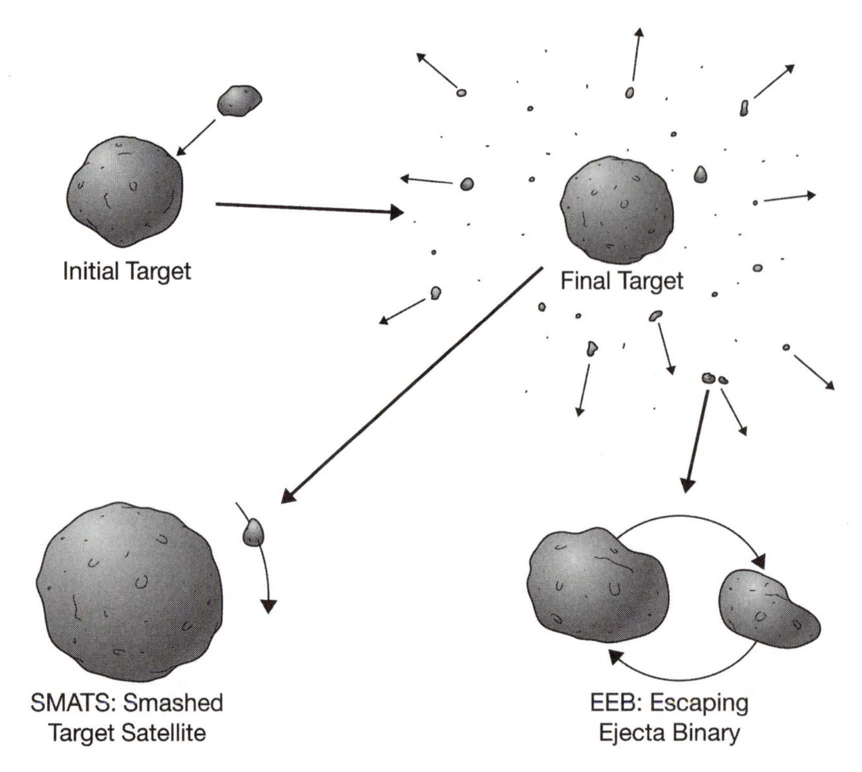

Figure 6.6 Scientists investigating the origin of asteroidal satellites use computer simulations to model large impacts, then can construct a movie of the results. The top left shows the situation before the impact, with a small object coming in from the right to hit a larger target. The top right portion of this figure is a still frame from the movie of the results, and shows a large amount of ejecta, or material thrown off from the impact, surrounding the target and moving away (in the direction of the smaller arrows), mostly escaping. The lower left and right show the two ways we expect asteroidal satellites to be formed. The lower left strips away most of the ejecta, showing only the central body and one piece of ejecta that remains in orbit as a satellite (its movement shown with a curved arrow). Finally, the bottom right focuses in on two pieces of ejecta that luckily ended up moving in a similar direction and became bound to one another as a binary system. These two satellite systems have very different properties: the escaping ejecta binary has a primary and secondary of nearly the same size, like near-Earth asteroid binaries are seen to have, while the size and orbit of the smashed target satellite is more similar to what is seen in the main asteroid belt. Illustration by Jeff Dixon.

orbit as a satellite (sometimes called "SMATS" for "smashed target satellite"). Another possibility is that two or more pieces of escaping mass could be ejected in the same direction, allowing them to become bound to one another into a single system (sometimes called "EEBs" for "Escaping Ejecta Binaries"). These scenarios are depicted in Figure 6.6.

The disruption of Comet Shoemaker-Levy 9 due to the gravitational tidal forces of Jupiter led to recognition of another possible way to create binary systems. Simulations of small bodies passing close to planets (at reasonable distances for NEO passes by Earth, Mars, or Venus) show that the gravity of the terrestrial planets can tear apart fragile NEOs, which may then reaccumulate into a binary system. However, an already-existing binary system making a close pass to a planet is likely to be completely disrupted. Further study is needed to see whether planetary close passes create more binary systems than they destroy.

The combination of rapid rotation, spherical primaries, and small sizes has suggested another mechanism for satellite creation among NEOs: spin-up and fission. Strengthless bodies (such as asteroid rubble piles are thought to be) will fly apart if their spin is too rapid and the centripetal acceleration at the surface is larger than the gravitational pull. Such objects would then lose mass from their surface as well as angular momentum, and the remaining mass would rotate more slowly. Some of the mass that was shed could remain in orbit as satellites. There are a few ways a NEO can be spun up—via close passes to planets and also by nongravitational forces like the "YORP" force, arising from the fact that objects take time to heat up and cool down and that irregular objects can heat unevenly (it is similar to the Yarkovsky Effect, described in Chapter 3). The rotation periods of most NEO binaries (including 1999 KW4) are close to their rotational limit, making fission an attractive explanation.

For the known TNO binaries, the explanations used to explain asteroidal systems do not result in systems that match observed properties. The wide separations appear to be best matched by the multi-body capture scenarios rejected for asteroid binaries. The objections to capture scenarios for asteroids are not as strong for the TNOs—the large TNO primaries have much stronger gravity than the relatively smaller asteroidal primaries. Furthermore, while the dynamical lifetimes of asteroidal systems are relatively short, TNO systems may be as old as the solar system itself. While the density of the present-day asteroid belt (as well as transneptunian space) may be too sparse to expect multi-body interactions, there was a much greater mass present as the solar system was forming. Given the larger number of bodies available for TNO interactions back then, it is thought that the observed number of binary systems could have formed and survived to the present day. The details of specific models await observations of smaller TNOs.

WHY NO COMETARY SATELLITES?

The previous discussion has all focused on satellites of asteroids or TNOs. To date, no comets have been confirmed to possess a satellite. Comet Hale-Bopp was suggested to be binary, but no firm observational evidence has confirmed it. Comets are not expected to have satellites, since the nongravitational forces on a comet (presumably acting on both primary and secondary) would reduce the stability of any system and tend to scatter the pieces. However, as noted earlier in the chapter, there were many good reasons to doubt the existence of asteroidal satellites until they were actually found.

SUMMARY

The sizes, shapes, and presence of companions for asteroids and TNOs vary considerably. The most straightforward way of determining these properties, direct imaging, only works for a small subset of targets at this time. Most sizes are determined using a combination of photometry in the visible and infrared spectral regions. Lightcurve studies have been used to determine the shapes of objects, as well as find satellites. Radar measurement, a form of direct imaging, is a powerful tool for determining shapes and finding satellites, though it is currently limited to nearby objects.

The formation of small body satellites occurred in different ways, depending upon where the objects are found. The TNO systems we see today are likely to have formed through capture via complex multi-body gravitational interactions near the start of solar system history. The large main-belt asteroids with satellites are thought to have gained those secondaries through collisions, either as ejecta fortuitously thrown into an orbit that could stabilize, or as multiple pieces of ejecta thrown off in the same direction became bound to one another. Near-Earth objects may have formed binary systems by fission as close passes to large planets or thermal forces increased their spin rate until they lost mass from their surface. A large fraction of NEOs are thought to be binaries (roughly 15 percent of all NEOs, and a much larger fraction of rapid rotators), while only a few percent of main-belt asteroids and TNOs have been observed to have satellites.

RECOMMENDED READING

Richardson, Derek, and Kevin Walsh. "Binary Minor Planets." In *Annual Review of Earth and Planetary Science.* Palo Alto, CA: Annual Reviews Press, 2006.

Warner, Brian. *A Practical Guide to Lightcurve Photometry and Analysis.* New York: Springer-Verlag, 2006.

WEB SITES

This Web site has a description of the first ground-based detection of an asteroidal satellite by the discovery team, along with movies and images: http://www.boulder.swri.edu/~merline/press_release.

This JPL radar research page contains a great amount of information on asteroid observations, including both more technical and more popular treatments, and many images, movies, and links: http://echo.jpl.nasa.gov.

7

Composition of Small Bodies

We know a great deal about the compositions of the asteroids, comets, and dwarf planets, despite the relative scarcity of material from them available to researchers. In this chapter we will discuss and describe the minerals commonly found on these objects and their distribution in the solar system, how scientists know about their presence, and how objects are classified on the basis of their compositions.

MINERALS AND ROCKS

On Earth, we are surrounded by rocks of various types: granite, basalt, marble, and sandstone, among others. These rocks are usually classified into one of three groups, depending on their origin: **igneous** rocks formed when magma or lava cooled and solidified, **sedimentary** rock formed from material transported by water or wind, and **metamorphic** rock, which formed from previously existing rock exposed to and changed by high heat and pressure.

Rocks are composed of groups of **minerals**, crystalline material sometimes made of one element, but most often compounds of several elements in set proportions. Common minerals include gypsum, talc, and calcite among many others. Some minerals are considered extremely valuable, such as diamond and sapphire, while others are exceedingly common like the quartz and magnetite that make up a large fraction of beach sand.

As discussed in Chapter 4 and 5, some elements are more commonly found in the solar system than others. In addition, due to reasons such as atomic size, charge, and the like, some elements are difficult to fit into crystal structures and are rarely found in minerals or are only found in a few minerals. In general, the most common minerals contain silicon, magnesium, oxygen, calcium, iron, and aluminum.

Quartz is the most common mineral at the Earth's surface, and is composed of silicon and oxygen, with a formula of SiO_2. Most of the minerals on Earth contain both silicon and oxygen, and are called **silicates.** The crystal structure of silicates varies widely, resulting in minerals of varying properties. Other important groups of minerals include **oxides** (which have oxygen, but not silicon) and **carbonates** (which contain carbon and oxygen in a specific arrangement).

ROCK-FORMING MINERALS

Two of the minerals found most commonly in Earth rocks (and in meteorites) are olivine and pyroxene. Olivine is a silicate with varying amounts of iron and magnesium. The pure magnesium version (or endmember) is called forsterite, with a formula Mg_2SiO_4. The pure iron endmember is called fayalite, and has the formula Fe_2SiO_4. In a crystal of olivine, small units of Mg_2SiO_4 or Fe_2SiO_4 share oxygen atoms in making up the large crystal. An atom of magnesium can share an oxygen atom with an atom of iron, meaning that olivine crystals can have any mix of iron and magnesium units. The composition of a crystal or larger sample of olivine is often described as the percentage of forsterite and fayalite: olivine with only magnesium and no iron is called Fo100Fa0 (or simply Fo100), while olivine with half iron and half magnesium is called Fo50Fa50 (or again, simply Fo50). Olivine is the most common mineral in the Earth's mantle, and it is also found throughout the solar system. Indeed, it has even been identified in dust around other stars and in-between stars.

Pyroxene is a more complicated mineral than olivine. Like olivine, it is a silicate that can contain magnesium and iron. However, its crystal structure can also allow other elements to be incorporated, usually calcium. The pure magnesium endmember is called **enstatite**, with a formula of $Mg_2Si_2O_6$. Enstatite is the major constituent of the enstatite chondrite meteorites and the related aubrite meteorites. As with olivine, there is also an iron-rich endmember $Fe_2Si_2O_6$ (ferrosilite), and pyroxene can be found with intermediate compositions between enstatite and ferrosilite. As mentioned, the crystal structure of pyroxene can also accommodate some calcium. However, because the size of a calcium ion is larger than magnesium or iron, there is only a limited amount of calcium that can fit into a pyroxene structure.

Another major mineral on the Earth, the Moon, and in meteorites, is feldspar. Instead of magnesium or iron, feldspar includes aluminum and either calcium, sodium, or potassium in addition to silicon and oxygen.

Although feldspar is commonly found in planetary and terrestrial samples, it is difficult to detect remotely, as further discussed in another section.

The final group of silicate minerals to discuss here are the **phyllosilicates**. These minerals include hydroxide (OH) or water in their crystal structure. Phyllosilicates on the Earth form from the chemical weathering of other minerals, and can include a number of elements such as magnesium, iron, aluminum, potassium, and calcium, in addition to silicon and iron. The importance of finding phyllosilicates is that their creation implies water either currently or at some time in the past. Phyllosilicates have been seen in meteorites, particularly the carbonaceous chondrites, and they have also been seen on Mars and some of the asteroids.

METAL

Metals are composed of a single element, and have specific properties that separate them from compounds. The set of atoms all share electrons, resulting in a greater ability to conduct electricity. Metals are relatively rare in rocks at the Earth's surface. The majority of the most common metals are extracted from ores, like iron or copper. Those metals that occur naturally, like gold and silver, are often considered quite valuable.

As discussed in Chapter 4, the cores of the terrestrial planets are composed of iron-nickel metal, and many elements that are chemically predisposed to be found in metal rather than rock have been brought into those cores as well. On the parent bodies of the chondritic meteorites, which are undifferentiated, metal still exists at the surface. The iron meteorites are the cores of disrupted and differentiated objects and other objects in which metal has collected due to impact heating. When meteorites fall to Earth, however, any iron metal present will quickly rust due to oxygen in the Earth's atmosphere.

ORGANIC MATERIALS

Geochemically, *organic* has a different meaning than what is found at the grocery store. An organic material is one that contains carbon, though it is often restricted to those materials that have carbon bonded to nitrogen or hydrogen. Much of the organic material on Earth was created by living things, and compounds like sugar, alcohol, and amino acids are all classified as organic.

However, organic material can also be created abiotically, without life. As noted in Chapter 4, carbonaceous chondrite meteorites have organic material in them, and it is also found on the surface of Saturn's satellite Titan. There is also evidence, discussed later, that it is present on comets. Organic material is also expected on other small body surfaces like main-belt and Trojan asteroids, and transneptunian objects. The organic compounds found in meteorites differ—the Murchison meteorite contains amino acids including many of those found in living things. The Tagish Lake meteorite, on the other hand, only contains simpler organic molecules. It is not clear at this point whether the organic material in Tagish Lake is more primitive (that is, organic material in Murchison started like that in Tagish Lake and further chemical reactions created amino acids), or whether the material in Murchison is more pristine (that is, Tagish Lake and Murchison had similar kinds of organic compounds, but much of what was in Tagish Lake was destroyed, leaving only what is seen today). It is also not clear how much some of these samples may have been changed and contaminated by their time on Earth, despite rapid collection and careful handling. Additional samples from organic-rich objects are dearly desired by geochemists, who hope to see a mission return material directly from an asteroid or comet, which would be truly pristine material.

The formation of organic compounds in early solar system history has been simulated in a number of experiments since the mid-twentieth century. The most famous is the **Miller-Urey experiment**, in which a mixture of water, ammonia, methane, and hydrogen was subjected to electricity simulating lightning. The resulting mixture included organic compounds including amino acids and sugars. The organic compounds present in meteorites and through the solar system most likely formed in similar ways through the reactions of these common molecules. The origin of life on Earth is a complex subject that has filled many textbooks and is still a source of ongoing research. The role that was played by small bodies in delivering organic material to Earth via impact is still not fully understood.

ICES

The elements most commonly found in the solar system are hydrogen, helium, carbon, nitrogen, oxygen, and neon, with far more hydrogen than the rest of the elements combined. These elements make a number of compounds with one another: water (H_2O), carbon dioxide and monoxide (CO_2 and CO), methane (CH_4), ammonia (NH_3), and hydrogen cyanide (HCN), among others. However, in the inner solar system the temperatures are too high to allow these compounds to be incorporated into rocks (and neon and helium are **noble gases**, which do not form compounds). As a result, rocks in the inner solar system are largely composed of other elements.

At greater distances from the Sun, temperatures drop and these compounds become stable. At outer solar system temperatures, water ice is as strong as rock, while volcanoes powered by warm carbon dioxide can spew out lava made of nitrogen (see Chapter 9).

Water ice can have several different crystal structures depending on temperature and pressure. The ice we encounter on Earth is crystalline, but we also find amorphous (or noncrystalline) ice in the outer solar system.

Amorphous ice is a sign of quickly cooled ice, which has remained at low temperatures since its formation.

REMOTE SENSING

For the vast majority of planetary studies, scientists have to gather data via telescope or satellite rather than from the surface of the object of interest. This is called **remote sensing**, as opposed to **in situ** (literally meaning "in the place" or "on-site") studies like those of the *Mars Exploration Rovers* or the *Huygens* probe to Titan. Compositional studies typically use reflected or emitted light from objects, although other means of getting compositional data are discussed at the end of the chapter.

THE COLORS OF LIGHT

Light is peculiar. In some ways and in some circumstances it is best described as being composed of massless particles called **photons** that carry energy. In other ways and circumstances it is best described as waves with a particular distance between crests (called the **wavelength**). The physics of light is a complicated subject, and for our purposes we must content ourselves with this abbreviated description.

The human eye can detect light with wavelengths from roughly 400–700 nm, which is called **visible light**. Visible light with short wavelengths (nearer 400 nm) is seen as blue; near the long wavelength end (700 nm) is seen as red. Light can be found far beyond these bounds, however. Light with wavelengths longer than 700 nm is called **infrared** or **IR** (beyond red), while **ultraviolet** or **UV** light has wavelengths shorter than 400 nm. Just as visible light can be detected by the eye, infrared light can be detected by skin as heat. Light can also be found beyond the infrared and ultraviolet ranges, as seen in Figure 7.1. However, the visible, infrared, and ultraviolet ranges are the most important for studies of asteroids, comets, and dwarf planets.

The wavelength of light and the energy it carries are connected via an inverse relationship: $E = hc/\lambda$, where E is the energy, λ is the wavelength, h is a constant (called Planck's constant), and c is the speed of light.

Astronomers can use the amount of light reflected from objects at varying wavelengths to measure their composition. This technique, called spectroscopy, is not limited to astronomical objects, but is also widely used on the Earth. The changing intensity of light coming from an object at different wavelengths is called a **spectrum** (plural: **spectra**).

MEASURING COMPOSITIONS VIA SPECTROSCOPY

Light interacts with matter in a particular way. An atom or molecule must do one of three things to any photon that hits them: **absorb** it (and absorb

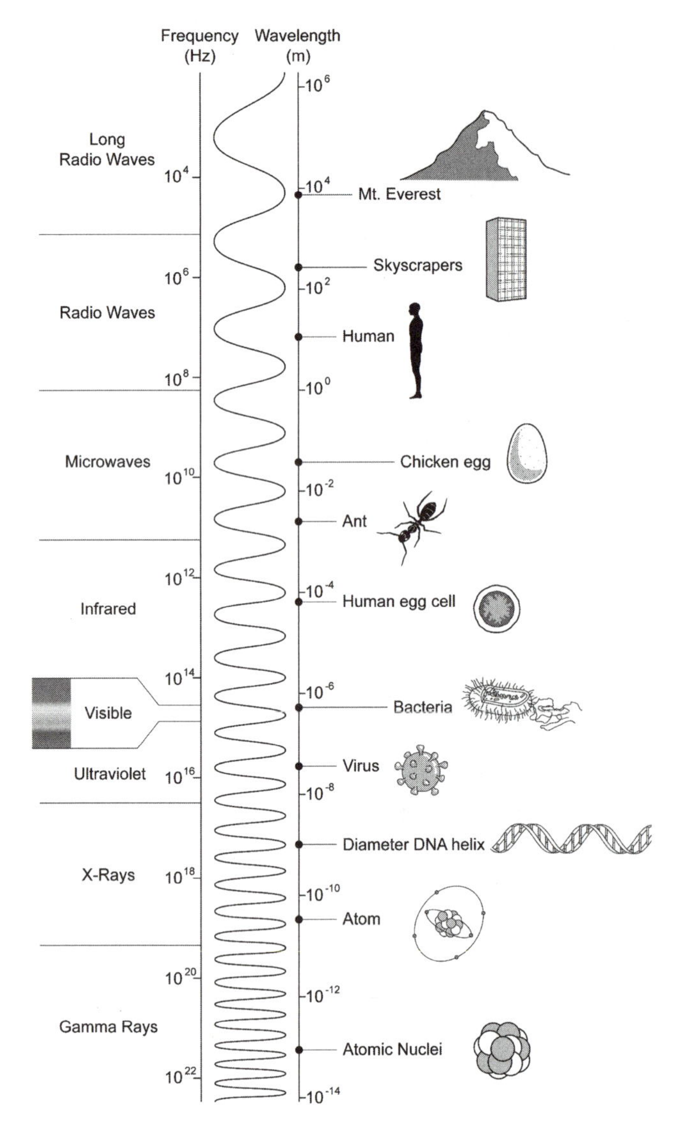

Figure 7.1 Light is divided up into several categories depending on its wavelength. Our eyes are only sensitive to a narrow range of wavelengths, called visible light (wavelengths near 0.5 μm, or 500 nm, or 0.000005 m). Shorter wavelength light carries more energy and includes ultraviolet (UV) light, X-rays and gamma rays. Longer wavelength light includes infrared (IR), microwave, and radio. Most compositional research for comets, asteroids, and dwarf planets is done using IR, visible, and UV light. Illustration by Jeff Dixon.

its energy), **reflect** it (changing the photon's path), or **transmit** it (allow it to pass through). Because of the specifics of atomic structure, only photons of a few specific energies can interact with and be absorbed by atoms of a particular element. Photons with different energies will not be absorbed. Similarly, if atoms are heated by absorbing photons, they can only cool down by giving off (or **emitting**) photons of certain energies. As mentioned before, the energy of a photon is related to its wavelength, and therefore the

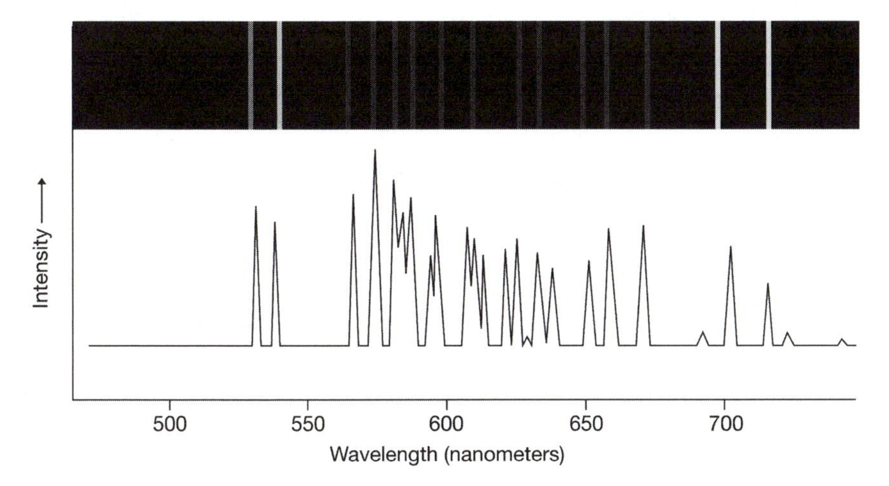

Figure 7.2 The atoms in a gas can only absorb and emit light of particular wavelengths. Seen here is the characteristic pattern of light emitted by neon atoms, familiar to us in neon lights. The top panel shows how the light from a neon lamp would appear after passing through a prism. The bottom panel is a graph of the amount of light as a function of wavelength, or a spectrum (in this case an emission spectrum). The wavelengths of emission can be easily read from the spectrum. This technique is at the heart of most compositional studies of asteroids, comets, and dwarf planets. Illustration by Jeff Dixon.

specific energies that are important for a given material mean that specific wavelengths are important for them. A familiar example of this phenomenon is seen in neon lights, whose characteristic color is a result of emission of photons that are mostly in the red part of the visible spectrum.

If we were to measure the amount of light coming from a neon lamp as a function of wavelength, we would come up with a graph like Figure 7.2—at most wavelengths, the lamp gives off no light, but at a few wavelengths it is bright. This particular set of lines is diagnostic of neon, and the gas in a neon lamp in space could be identified if its spectrum were seen. Each element has a particular combination of photon energies (and thus wavelengths) that it produces in a spectrum, making remote identification possible. In fact, helium was first detected on the Sun by astronomers before it was found on Earth. Because its spectrum was like no terrestrial material then known, it was named for the Sun (*helios* in Greek).

Within a decade of the first stellar spectroscopy, the spectra of comets were being measured for the first time. By the late 1870s, carbon compounds were detected in comets. Spectroscopy has continued to be an important tool for cometary studies since those first spectral measurements.

The coma and tail of a comet are observed to have emission lines, like the neon lamps discussed before. Some light is also solar light reflected by the dust grains. The intensity of this reflected light changes slowly and smoothly with wavelength, and is called the spectral **continuum**. The amount of absorbed and emitted light is usually measured in comparison to the continuum.

ALBEDOS

The fraction of light reflected from an astronomical object compared to the light falling on it is defined as its albedo. The most commonly used albedo values are those measured in the visible wavelength range. Although not technically compositional measures, albedo values can be useful guides to composition. As a rule of thumb, more organic-rich and carbon-rich bodies tend to have low albedos (in the range of 5–10 percent, usually expressed as a decimal range (0.05–0.10), with cometary nuclei generally having albedos of 0.04. Silicate objects of ordinary chondrite-like compositions have albedos in the range of 0.2–0.4, as do metallic objects. Ice-covered objects will have higher albedos still, reflecting 60–80 percent or more of the light that falls on them.

The albedo of an object is also an important factor in its surface temperature. As already mentioned, photons hitting a surface will either be reflected, transmitted, or absorbed. A high-albedo object will reflect most of the light that hits it, leaving very little to be absorbed, while the opposite is true of a low-albedo object. Light absorbed by an object will heat it up, as seen every day here on Earth when things are left in the sunshine. Many of us also have first-hand experience with the idea that dark objects heat up more than lighter ones, as anyone who has walked barefoot across both blacktop and sidewalks will agree!

Perhaps surprisingly, all objects emit some light. The amount and average energy of the light is strongly related to the temperature of the object. This temperature-dependent emission, called **blackbody radiation,** is not dependent upon composition, except to the degree that different compositions have different albedos. By looking at the distribution of light across many wavelengths, the temperature of an object can be measured. This is how we know the temperature of the Sun, for instance. Asteroids, comets, and dwarf planets all emit light in the infrared part of the spectrum, and measurements of their temperatures are relatively common. By measuring their brightness at visible wavelengths at the same time, however, albedos can be determined. This is because both the visible brightness (dependent upon the amount of sunlight reflected) and the infrared brightness (dependent upon the temperature) are dependent upon the object's albedo. This gives sufficient information to allow the albedo to be determined.

For some objects, albedo measurements are all that we have. In these cases, scientists can estimate likely compositions based on the albedo measurement. In some cases, the estimated composition may be incorrect, but often such estimates are "close enough." For objects of particular importance, however, scientists are rarely satisfied with albedo measurements alone.

GAS PHASE SPECTROSCOPY

As an introduction, let us consider what may provide a familiar example: light coming through the earth's atmosphere. Our atmosphere is mostly composed of nitrogen gas, with a smaller amount of oxygen gas and a little bit of water vapor, carbon dioxide, methane, and other gases. We know from experience that most visible light passes through the atmosphere unimpeded (at least on a clear day) and from this we do not expect any of these gases to have strong absorptions in the visible region. In the UV and IR, however, the atmosphere can be strongly absorbing. In some cases, this has had a direct effect on humankind—ozone (O_3) absorbs UV light at particular wavelengths, preventing much of it from reaching the surface. The "ozone holes" near the Earth's poles are regions of depleted ozone, which allow increased amounts of UV to reach the surface, resulting in increased sunburn and skin cancer. Absorptions in the IR are also important—water, carbon dioxide, and methane all absorb IR photons, while they are transparent to visible light. Visible light from the Sun passes through the Earth's atmosphere and warms the ground, which reemits IR photons. These IR photons are then absorbed by the water, carbon dioxide, and methane in the Earth's atmosphere, leading to an overall greater temperature (this is the well-known greenhouse effect).

The amount of gas seen by an observer is often measured in terms of its **optical depth**. This is a unitless number that represents the amount of material required to scatter or absorb roughly one-third (actually $1/e$ or 1/2.718) of the light that falls on it. The mass of gas corresponding to one optical depth is dependent upon the specific molecule and wavelength in question. Spectroscopy of gas molecules has been important in the study of planets for over a century. It is still the best way we have to understand the composition of the giant planets. Small bodies are more poorly endowed with gases than planets, but they are still an important component of comets. The gases in a cometary coma are observed when they emit photons, like the neon lamp used as a previous example. These **emission spectra** have contributions from the presence of many different compounds.

For many molecules in a comet, so-called **fragment species**, which are component parts of a **parent molecule**, are observed instead of or in addition to the original molecule. For instance, most cometary activity is driven by the sublimation of water ice, and therefore we expect to find evidence of water. Once in space rather than on the surface of the comet, water molecules (H_2O) can be broken up by high-energy photons into OH and H, or H_2 and O, all of which can be broken down even further. Alternately, the water can lose an electron and be ionized, symbolized as H_2O^+. All of these breakup products have different spectral signatures, and by observing all of them, scientists can better determine the rate at which water is leaving the comet.

The primary fragment species that have been seen in comets are various combinations of hydrogen, oxygen, nitrogen, and carbon. These are thought to be derived from the breakup of water, ammonia (NH_3), carbon dioxide (CO_2), carbon monoxide (CO), and various organic molecules. Sulfur has also been detected in comets spectroscopically. However, the parent molecule for the sulfur has not been clearly identified. Because fragment species are often what is observed, and each parent molecule can produce several fragment species, any suggested parent for the sulfur will also affect the expected amount of CH, for instance. At this point, the two most likely suspects appear to be H_2S (hydrogen sulfide—the gas that gives rotten eggs, among other things, their distinctive smell) and OCS (carbonyl sulfide—another unpleasant-smelling gas that is the main sulfur compound in the Earth's atmosphere). Complicated organic molecules can be broken up in several different ways, with the fragments named by the number of carbon atoms present—C_2 fragments have two, while C_3 fragments have three.

In addition to detecting different molecules, spectroscopy can be used to measure the relative abundances of different isotopes, such as hydrogen and deuterium. This measure, called the D/H ratio, is characteristic for each body, as discussed further in Chapter 5. With the proper equipment, differences can also be seen between two molecules of the same composition that are structured slightly differently. For instance, two different forms of water (called ortho and para) can be distinguished. The relative amounts of ortho and para water in ice is temperature-dependent, and spectroscopic observations of the ortho to para ratio in comets have been used to estimate their formation temperatures—typically 25–40 K, the temperatures found beyond Neptune.

SOLID SURFACES

Spectroscopy of solid surfaces is much more complicated than spectroscopy of gases. The absorption, scattering, and transmission of photons is dependent upon the particle size and temperature of the surface, as well as the angle of the incoming light and other factors. Typically, light can only penetrate into rock or mineral grains a distance roughly equal to a few times its wavelength—only a few micrometers for visible and infrared light, and up to a few meters for radio observations and radar studies. Most cometary spectra have some contribution from the coma, which greatly complicates analysis of their surfaces. As a result, the majority of work in solid surface spectroscopy has focused on asteroids and transneptunian objects.

Unlike free-flying molecules of gas, solids are usually organized into structures (or crystal lattices). Atoms in a crystal lattice have some ability to move within the lattice, but are for the most part confined to a small area. Nevertheless, the ability to move a bit changes the absorptions seen in solids from the narrow lines seen in gases to broader bands covering a range of wavelengths. The wavelength range for a solid depends on composition, so olivine with different amounts of fayalite can be distinguished, for instance. This has led to the **band center** being an important parameter measured by spectroscopists. In addition, the **band depth** is also an important measure. Figure 7.3 shows schematically how these parameters are measured.

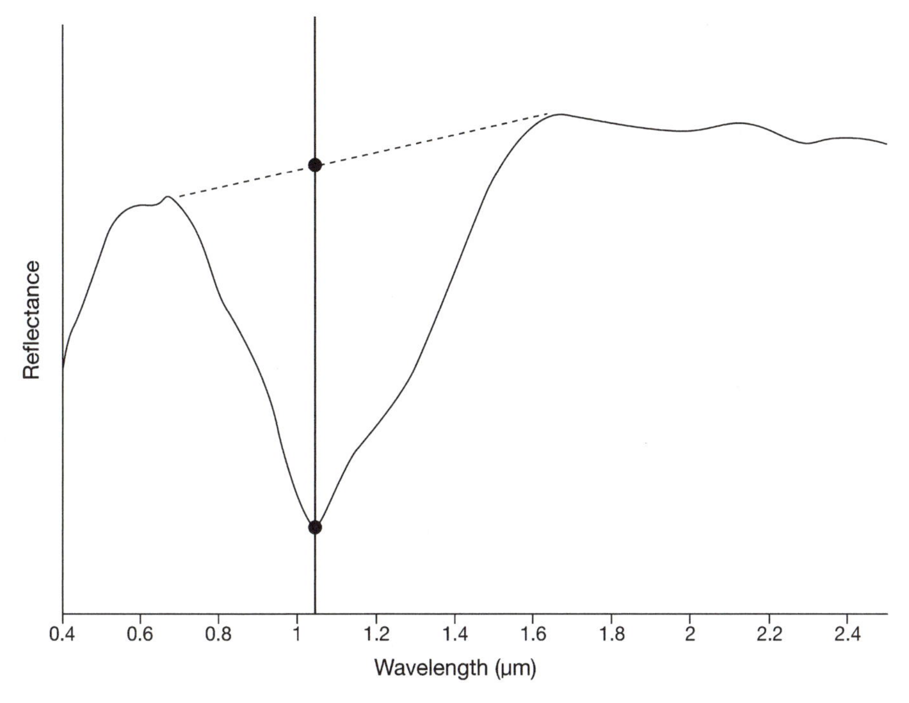

Figure 7.3 Scientists remotely measuring the compositions of small bodies typically use parameters read from reflectance spectra. Here, several of the more important concepts are shown using a spectrum of olivine. The band center is the vertical solid line, which is drawn at the wavelength of minimum reflectance (roughly 1.05 μm in this spectrum). The continuum is an estimate of how the spectrum would look without the absorption band and is here shown with a dashed line. The band depth is the difference in reflectance between the continuum value and the value at the band minimum, those points here shown with solid points. Illustration by Jeff Dixon.

By measuring the spectra of minerals and meteorites in laboratories on Earth, spectroscopists have made great progress in tying the spectra of the small solar system bodies to specific compositions. However, real objects are mixtures of many minerals, not just one, some of which are spectrally featureless. Different types of mixtures require different analytical approaches. The simplest type of mixture is an **areal** or **linear** (sometimes called "checkerboard") mixture. In an areal mixture, the different spectral types are separated from one another—a simplified view of the Earth is an example. The spectrum of the Earth from far away has a contribution from the oceans and a contribution from the continents (for the moment we will pretend the continents have a single spectrum, ignoring different rock types and vegetation as well as ignoring the atmosphere). A photon reflected off of the Earth will only encounter one type of material (ocean or continent) before coming to our telescope. Even if it is reflected more than once before heading in our direction, it is not likely to bounce off of both ocean and continent. In areal mixes, the spectrum that is measured is a simple combination of the different spectra in the ratio of the different areas; in the case of the previous example, the average Earth spectrum would be 0.71 times the ocean's spectrum plus 0.29 times the continental spectrum since the Earth is 71 percent ocean and 29 percent continent.

The second, and more complicated type of mixture is an **intimate** (or **nonlinear**) mixture. A potentially familiar example of an intimate mixture is typical beach sand—a variety of grains of different colors. From a distance, the different grains cannot be distinguished from one another, and a beach appears relatively uniform. In an intimate mixture, photons typically are reflected from more than one material before they are measured. Therefore, there is no simple way to determine the fraction of each contributing spectrum. However, sophisticated theoretical models have been used to estimate the composition of intimate mixtures from analysis of observed spectra.

Most of the minerals that make up asteroid and TNO surfaces have distinctive absorption bands in the visible and near-infrared (NIR) spectral regions, particularly between 1,000 and 5,000 nm (1–5 μm). This includes almost all of the mineral types mentioned before: olivine, pyroxene, and other silicates, organic compounds, water ice, and other ices. For this reason, most compositional studies take advantage of this spectral region.

However, there has been an increasing amount of research that focuses on the mid-infrared, usually defined as the 5–25 μm spectral region. Silicates and ices also have spectral features in this region, and several spacecraft sent to Mars have brought along instruments sensitive to the mid-IR. This wavelength region is very sensitive to temperature, and most asteroids have large amounts of emitted light within this wavelength range, which can complicate analysis. In addition, the surfaces of small bodies often are covered in particles with sizes of 5–25 μm. This also affects the analysis because the way light is reflected and emitted from particles changes when their size is close to the wavelength of light. However, mid-IR observations are of increasing interest, and the Spitzer Space Telescope has provided a large amount of mid-IR data for asteroids and comets, which is helping to spur further research.

SPACE WEATHERING

The processes that occur in the regolith of small bodies discussed in Chapter 8, collectively called **space weathering**, are thought to change their spectra, potentially confusing their interpretation. Studies of lunar material suggest that micrometeorite impacts vaporize silicate particles in the regolith of airless bodies, with some of the iron in those silicates redepositing on other regolith particles. This process changes the spectrum of the regolith of rocky bodies from a purely silicate one to a mixture of silicate and metal, and in extreme cases it can result in spectra that look like mostly metallic surfaces. This change can make it more difficult to interpret the surfaces of space-weathered objects.

The majority of space-weathering research has focused on the Moon and inner-belt asteroids. The processes that cause space weathering on these objects should also be occurring with varying strength on all objects in the solar system that lack atmospheres. The possible spectral changes are difficult to detect on outer-belt asteroids, whose spectra are relatively featureless to begin with.

However, there is some evidence that the spectral effects of space weathering can be seen in Centaurs and transneptunian objects. Experiments simulating the exposure of ice to the solar wind and UV light from the Sun show that the high albedo of the ices found in the outer solar system tend to become darker, consistent with existing data. Observations of TNOs are still few enough that firm conclusions are still in the future.

CLASSIFYING SPECTRA: TAXONOMIES

The first surveys of asteroid spectra were done in the 1970s and used visible data only. Astronomers quickly noticed similarities between some asteroid spectra and meteorite spectra. They separated the spectra into three groups and somewhat optimistically gave them names reminiscent of the meteorites they resembled: S for "stony," C for "carbonaceous," and M for "metal." All taxonomies that have followed have tried to keep similar names as the initial taxonomy, which has allowed for some continuity but also occasionally has created some confusion.

Most of the letters of the English alphabet have now been assigned to an asteroid spectral class. However, there are still three main groups. The **S complex** comprises asteroids with visible-wavelength spectra shaped like an upside-down U—absorptions due to iron oxide at the short wavelengths and silicates at the long wavelength create that shape. S-complex asteroids include Ida, Eros, Apophis, and Juno. The **C complex** includes asteroids with relatively flat spectra, save for the iron oxide absorption at the short wavelengths. These asteroids tend to have low albedos, and include Ceres, Pallas, and Mathilde. The **X complex** includes the M asteroids. These objects all have visible spectra that are featureless, save for slope differences between them. The albedos of X-complex asteroids vary, with some taxonomic schemes including the albedo in their classifications. The C, S, and X complexes are further broken down into classes, including (somewhat confusingly) classes named C, S, and X.

There are also some asteroid classes that don't fit in any of the main three complexes. The two of note are the **V class** and the **D class**. The V-class asteroids have visible spectra like Vesta, with a silicate absorption much deeper than the S-complex objects. D-class asteroids have steep, featureless spectra, with slopes much steeper than the X-complex asteroids. Many Trojan asteroids have D-class spectra, which are also similar to the spectra of comet nuclei and the Martian satellites Phobos and Deimos. Figure 7.4 shows example spectra of the major asteroid classes.

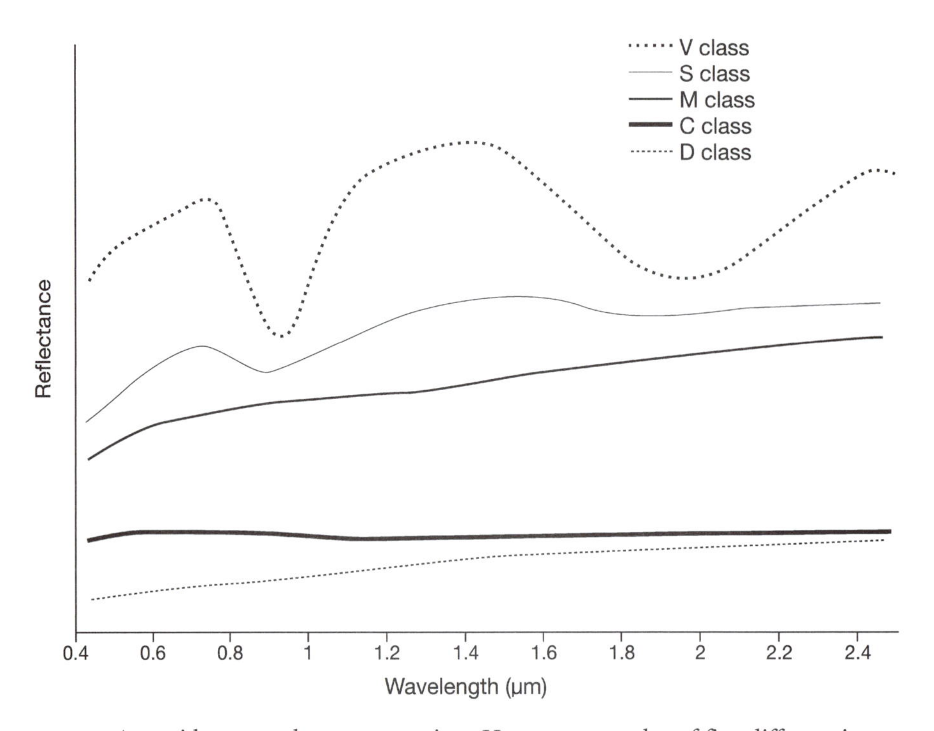

Figure 7.4 Asteroid spectra show great variety. Here are examples of five different important asteroid spectral classes. The D class has low reflectance, but gets steadily brighter at longer wavelengths. Many comets and outer-belt asteroids have D-class spectra. The M class has a similar spectrum, though with higher reflectance. These objects are associated with iron meteorites, though several other compositions are possible. When the absolute brightness is taken into account, D-class asteroids have spectra much more steeply sloped than the M asteroids. C-class asteroids are intermediate in brightness but have flat spectra. The largest objects in the asteroid belt are C-class, including Ceres and Pallas, and carbonaceous chondrite meteorites have similar spectra. The final two classes shown both have absorptions near 1- and 2-μm, indicating the silicate minerals olivine and pyroxene. S-class asteroids look similar to ordinary chondrites, though there are some differences in spectral slope. The S asteroids dominate the near-Earth asteroid population and include Ida, Gaspra, Eros, and Itokawa. Finally, the V-class asteroids are named for Vesta. Their spectra are the same as those of the HED meteorites and have been used as evidence that those meteorites come from Vesta. They reflect much more of the light that falls on them than the other major classes. Illustration by Jeff Dixon.

FROM TAXONOMY TO COMPOSITION

As more and more data become available, it is increasingly clear that there can be wide variations in composition within a single asteroid spectral class. It is also clear that different classes can have very similar compositions. Nevertheless, while the use of taxonomies runs the risk of oversimplifying matters, they also serve as a convenient shorthand if understood as a tool for guiding further work.

To move beyond taxonomy, planetary scientists have sought means of converting reflectance spectra into quantitative measures of composition.

Through the late-twentieth century, minerals were studied in the laboratory to understand how their structures and compositions lead to their spectra. These studies continue today. They have generally found that as compositions change, the structure of a mineral changes, leading to shifts in band center and band depth.

THE S ASTEROID PROBLEM

All these factors come together when using reflectance spectroscopy to determine the composition of asteroids. The first asteroidal spectrum to be obtained was that of Vesta, which was quickly seen to be similar to the HED meteorites. Astronomers in the 1970s and 1980s were confident that all meteorite types would be easily associated with asteroids or asteroid classes. One of the associations investigated was that of the ordinary chondrite meteorites, the most common meteorites seen to fall to Earth (see Chapter 4), and the S-class asteroids, the most common asteroid class in the inner asteroid belt. The ordinary chondrites are of interest not only because they are common, but also because they represent material that is largely unchanged since the formation of the planets. Determining their distribution in the asteroid belt sheds light on conditions 4.5 billion years ago and provides insight into how the inner planets were put together.

At the first spectra for the S-class asteroids were analyzed, they were seen to have similar features to the ordinary chondrites: absorptions due to olivine and pyroxene, and roughly the correct albedos. However, as planetary astronomers looked at S asteroids in more detail, they realized that there were mismatches with ordinary chondrites that were potentially quite important. The two most obvious discrepancies were that the S asteroids had shallower absorption bands and also steeper spectral slopes. These were exactly the changes seen in lunar samples when they suffer space weathering, leading some to propose that the S asteroids were simply space-weathered ordinary chondrites. Others pointed to the very different conditions present at the Moon versus the asteroids and the lack of evidence for space weathering in the meteorites, and doubted that space weathering was occurring on asteroids to any degree. They proposed instead that the S asteroids known at the time had all melted early in their history and weren't ordinary chondritic at all. In this view, the ordinary chondrites were only found at very small sizes. The difference between these interpretations of S-asteroid composition led to very different conclusions about the early solar system. The question of what the S asteroids are made of, and where in the asteroid belt the ordinary chondrites originate, was so central to asteroid studies that solving the "S-asteroid problem" was a major goal of both the *NEAR Shoemaker* and *Hayabusa* missions. While agreement is not unanimous, most astronomers have become convinced that the space weathering scenario is correct.

TNO TAXONOMIES

Transneptunian objects are generally much fainter than asteroids, and as a result it is much more difficult to obtain spectral information. A taxonomic classification system for TNOs has only recently been developed, with four classes: BB, BR, IR, and RR. These classes describe the visible and near-IR spectrum using four wavelengths from 0.4–1.25 μm. The class names are inspired by their spectral shapes: RR objects are the reddest (their reflectance *increases* most rapidly with increasing wavelength), while the BB objects are the bluest (their reflectance *decreases* with increasing wavelength, *or* in this case increases *least rapidly* with increasing wavelength). The BR and IR classes are intermediate in redness compared to the BB and RR classes. A limitation of this taxonomic scheme is that it is designed for those TNOs that are featureless, and objects like Pluto and Eris are difficult to classify.

Detailed spectra have only been obtained for a few TNOs. However, spectral mixing models of the sort described for asteroids have also been calculated for TNOs. Modelers have shown that the data for the four TNO taxonomic classes can be matched using differing amounts of organic material, water ice, and minerals. The differences between the spectra seem to largely depend upon the amount of organic material, though the amount and effect of space weathering on TNO surfaces is not well known, and could play a major role.

Independently of the TNO taxonomy described here, researchers have been incorporating the more detailed spectra into a still-forming classification scheme based on composition. This scheme also has four rough categories, splitting TNOs into methane-rich, featureless, and water ice-rich classes. The water ice-rich class is further divided into objects with crystalline water ice and amorphous water ice, indicative of more-heated and less-heated objects.

COMET TAXONOMIES

A taxonomy scheme for comets has been much more elusive than for asteroids or even KBOs. Cometary nuclei are rarely observed spectroscopically because of contributions from the coma. What nuclear spectra do exist are consistent with D-class or C-complex asteroid spectra. Because most cometary spectra are emission spectra from the coma, this has been the focus of classification attempts. Unlike asteroidal spectra, which are consistent throughout an orbit, cometary emission spectra change significantly with solar distance and temperature. As a result, the spectra themselves cannot be used for classification. However, those spectra can be used to calculate compositions including the relative amounts of water and organic material on an object. It is these compositions that have been used for classification schemes.

The most promising basis for a cometary taxonomy seems to be the abundance of the C_2 and C_3 fragments relative to CN and OH. The comets that originate in the Kuiper belt have less of these fragments than those that originate in the Oort cloud. This has been interpreted as evidence of different abundances of organic material in the formation locations for those bodies, consistent with current theories. However, in contrast to the hundreds of asteroids that have been observed and classified, there are relatively few comets for which spectra are available.

OTHER COMPOSITIONAL INFORMATION

While much of what we know about small body compositions comes from spectroscopy, scientists can get indirect information about composition by other means. The most obvious is by studying actual samples in earthbound laboratories. Meteorites have provided the vast majority of detailed information about asteroid compositions that we have, as discussed in full in Chapter 4. Cometary samples have been returned by the *Stardust* spacecraft (see Chapter 12), and those samples are currently under analysis. Sample returns from asteroids are in progress, though it is not certain if they will be successful.

Density information is available for a handful of objects. The densities of ice, rock, and metal are quite different from one another, a fact that has been used in some cases to make estimates of composition when only density information is available. This has been most useful when high densities are seen, which is a sign of a metallic composition. Lower densities can be interpreted as due to the presence of ice, or alternately as evidence of high porosity and many fractures and cracks in the object, as discussed further in Chapter 9.

There are other indirect measures of composition. One is observations via radar. Metal is featureless in the visible and near-infrared spectral regions, and because radar is sensitive to metal content, it has been the most definitive means of identifying metal-rich objects including apparently all-iron bodies. Another indirect means is interpretation of thermal properties. Metal, ice, and rock heat up and cool down at different rates, a property called thermal inertia. Measurements of thermal inertia, though difficult, can give some insight into the composition of an object's surface.

SUMMARY

The minerals that make up the asteroids, comets, and dwarf planets are largely the same minerals that are found on Earth. However, they are found in different proportions to one another, and can be found with different compositions than seen on Earth. While a few samples of small bodies have been retrieved by spacecraft, and meteorites are pieces of additional objects (although the exact objects are usually unknown), the vast majority of compositional data we have is derived from remote sensing. Spectroscopy is the most commonly used remote sensing technique for determining composition, and it has been used to detect silicates, ices, organic compounds, and others. Gas-phase spectroscopy has been used to study the coma and tails of comets and to conclude that a variety of organic compounds and ices exist on cometary nuclei. The vast amount of spectroscopic data available for the surfaces of asteroids has necessarily led to classification schemes to handle the data, which can provide further insight, including suggestions that some processes like micrometeorite impact or exposure to the solar wind can change spectra. Classification schemes for cometary and TNO spectra have been proposed, but the relative rarity of those data mean that these schemes are still in their infancy.

WEB SITES

Northern Arizona University hosts this Web site that provides further information about the minerals found in meteorites, including chemical formulas and appearance: http://www4.nau.edu/meteorite/Meteorite/Book-Minerals.html.

A general introduction to the electromagnetic spectrum and its different regions is on this NASA-owned Web site: http://imagine.gsfc.nasa.gov/docs/science/know_l1/emspectrum.html.

"The Basics of Light" contains additional information about the electromagnetic spectrum: http://fuse.pha.jhu.edu/~wpb/spectroscopy/basics.html.

This site maintained by the United States Geological Survey has a detailed explanation of a wide variety of spectroscopic techniques at a higher technical level than presented in this book: http://speclab.cr.usgs.gov/PAPERS.refl-mrs/refl4.html.

The USGS provides a library of reflectance spectra for minerals at this Web site: http://speclab.cr.usgs.gov/spectral-lib.html.

A large amount of remote sensing data is available for small bodies. This Web site is an excellent resource for asteroid spectra from the Small Main Belt Asteroid Spectral Survey (SMASS): http://smass.mit.edu.

NASA maintains a central clearinghouse for a wide variety of remote sensing data from both missions and earthbased research at this Web site: http://pdssbn.astro.umd.edu.

8

Surface Processes

We are famously warned by our parents and teachers not to "judge a book by its cover." But what if the book is unopenable and the cover is all we can see? Or we have a few paragraphs, but don't know for certain which book contains them? Planetary scientists have samples of some bodies (as seen in Chapter 4), which have given us great insight into the formation and evolution of the asteroidal and cometary populations as a whole. We have been able to tease out some information about small body interiors, as we will discuss in Chapter 9. However, the bulk of our knowledge of the small bodies and dwarf planets has been obtained by remote sensing, which is only sensitive to the top few meters (or, depending on technique, even less!) of their surfaces.

The surfaces of the asteroids, comets, and dwarf planets are still changing today. They are exposed to impacts both large and very small. They are exposed to the solar wind and harsh ultraviolet light. These changes have implications for our understanding and interpretations of small bodies and how what we learn from meteorites and IDPs can be best applied. In this chapter we will look at small body surfaces and their ongoing evolution.

IMPACTS

The small bodies do not experience rain or wind, they do not have volcanoes that erupt lava (at least in the last several billion years). However, they do experience impacts just as the larger planets do. These impacts come in all sizes, from tiny ones with dust grains whose individual effects are tiny but cumulative effects are large, all the way to catastrophic collisions that disrupt objects and send the pieces throughout the asteroid belt, and even eventually to impact the Sun or planets. Small impacts are more common than large impacts, but there have been billions of years for impacts of all sizes to occur.

The surfaces of small bodies are covered with craters as a result of these impacts. The average collisional speed experienced by objects in the asteroid belt is roughly 5 km/s, resulting in shockwaves upon impact. These shockwaves transfer energy very effectively into the target material, pulverizing solid rock into pebbles, gravel, and dust and throwing it out from the impact site, leaving a crater behind. This pulverized material, called ejecta (since it is ejected from the impact site), will land nearby, fly for great distances, or escape the body altogether depending on the speed of the impact, the density of the impacted rock, and the strength of the body's gravity. After billions of years of impacts, the surface of a small body is expected to be covered in the ejecta of uncounted impacts, with a powdery or gravelly texture. If one dug through this layer, one would eventually reach solid rock (at least on larger bodies; a further discussion of thoroughly fractured bodies is found in Chapter 9). This layer of broken-up rock is called regolith. Regolith is found on larger airless bodies like the Moon, as well as small bodies, as demonstrated in Figure 8.1.

The shockwaves during impacts that are so effective at creating ejecta are also effective at destroying the impacting material. Most of the meteorites that we find on Earth originated as smaller objects that broke up in the

Figure 8.1 The surfaces of most small bodies is covered in a powdery dust called regolith. The regolith is composed of broken-up fragments of rock created during impacts. In addition to asteroids, it is seen on the surfaces of other airless bodies like the Moon. This image of an astronaut's footprint on the Moon shows the consistency of the regolith. NASA.

atmosphere and fell relatively slowly. The objects that are large enough to make it to the ground intact and create craters are almost entirely vaporized themselves in the impact. The same is thought to be true for the impactors on the asteroids, comets, and dwarf planets; scientists do not think that there is a large amount of mixing between the impactor and the target, and current theories suppose that the vast majority of material in a regolith originated on the target body itself.

CRATERING AND CRATER COUNTS

At the most basic level, cratering can alter the shape of an object. Vesta, for example, has a giant crater near its south pole, which is visible in Hubble Space Telescope images. This large crater is thought to have been formed in the same impact whose ejecta created the Vesta dynamical family, some of which eventually reached Earth as the HED meteorites, as was discussed further in Chapter 4. The asteroid Mathilde also can be seen as obviously shaped by impacts, as all of its contours appear to be the edges of craters.

For the most part, the number of craters on a surface increases with time. After a flurry of impacts very early in solar system history, the frequency of impacts has evened out. Knowing the rate of impacts and counting the number of craters on a body allows its rough age to be calculated. The exact age is dependent upon a large number of factors that have been difficult to quantify. For instance, we need to know how often objects of different sizes hit a body. There are fewer large objects than small ones, but how the relative proportions of large and small bodies change for different places in the solar system (or whether it is the same everywhere) is a matter of some debate. The distribution of impacting bodies for a given body is called the **production function** since it is the set of objects that produce craters. While it is not clear how much the production function differs from body to body, it is clear that it changes with time for a given body—if nothing else, there were more impacts at all sizes early in solar system history, with a steady decline for billions of years.

We might imagine that the number of craters will grow with time. However, at some point enough craters have collected on a surface that any additional craters must fall on top of a crater that's already there. This situation is called **equilibrium saturation**, and represents a practical limit to crater counting. However, equilibrium saturation is reached at different times for different size craters—large impacts are rare enough that they only accumulate slowly. Small craters will degrade larger craters that they overlap, but the large crater is usually still identifiable.

Objects like the Earth and Io are quite clearly *not* in equilibrium saturation at any size range. These bodies have erased most of their craters through geological activity: Io through relentless volcanic action, the Earth mostly by erosion via wind and water. Other objects, like Mars, Venus, and the Moon, have some areas with high crater densities, and others with lower densities. This suggests that geological activity was active at different times in different areas. The data we have for asteroids suggest that they are close to crater saturation, though the images of comets suggest a complex history, discussed further in the following section.

Some of the material thrown from craters is seen as boulders on surfaces. With the high-resolution images available for Ida, Eros, and Itokawa, there have been studies done of the size distribution and location of boulders. On Eros, the distribution is consistent with their creation in the most recent impact crater, named Shoemaker. On Itokawa, however, the boulders appear to be too big to have been made in any of the impacts on its surface! This has been used to argue that Itokawa is a disrupted and re-accumulated **rubble pile**, and that the boulders formed when Itokawa was part of a larger body.

FAMILY AFFAIRS

The collisions that make craters and create regolith are sometimes large enough to generate ejecta blocks that are kilometers in size. Large enough impacts can disrupt a body entirely. After a so-called **catastrophic collision**, the largest remaining piece of the target is only half the size of the original object or smaller.

These large impacts have effects that can still be seen billions of years later. Because the ejecta and fragments of the original object had relative speeds after the collision that were small compared to their orbital speed around the Sun (though of course *much* larger than the escape speed), their post-impact orbits remain similar. Groups of objects with similar orbits have been identified and interpreted as **collisional families** or **dynamical families**. These include some very large groups, including the Koronis, Themis, and Eos families. Dynamical families are named after their lowest-numbered member. Family studies are a popular topic in asteroid science, as they potentially provide a means of probing the internal structure of an object and can also be used to determine the frequency and likelihood of collisions early in solar system history. While large families are all quite old, small families caused by the disruption of smaller objects are still being created. One grouping, the Karin cluster, was the result of an impact only a few million years ago—a mere blink of an eye in comparison to the 4.5-billion-year age of the solar system.

REGOLITH

The formation of regolith takes time. As noted, the amount of ejecta retained by a body depends in part on its size and the strength of its gravity. On the smallest bodies with the weakest gravity, the vast majority of ejecta moves much faster than escape velocity and regolith formation is impeded or absent. On larger objects, most ejecta is retained. As a general rule of thumb, the speed of material in ejecta depends upon the size of the particles. Boulders move more slowly than hand-sized rocks, and both of those types move more slowly than sand-sized particles, with dust moving most rapidly. As a result, and in combination with the dependence of regolith retention on ejecta speed mentioned previously, the particle sizes found in regolith are expected to differ on objects of different size.

In studying the crater populations on the asteroids Ida, Gaspra, and Eros, it was found that there were fewer small craters than expected given the number of larger craters. It seemed as though the smaller craters were being erased preferentially from these asteroid surfaces. On Earth, large craters are recognizable for long periods of time because smaller craters can be eroded relatively quickly (or at least "relatively quickly" compared to the age of the planet). Without wind and water, erosion as we know it on the Earth does not occur on small body surfaces. However, there is still a very slow change in the appearance of surfaces as features like craters become less sharp with time. This is largely due to the creation of regolith and micrometeorite impacts.

These gradual, steady changes should not act to erase small craters, however. One group of scientists realized that the answer might come from larger, infrequent impacts. Impacts create seismic waves just like earthquakes, which can affect the local area. According to computer modeling of impacts, the seismic waves shake the regolith, smoothing it out to some degree. Because the smallest craters are the shallowest craters, they are the easiest to erase, with even relatively small impacts able to create sufficiently large seismic waves.

Looking at the craters still present on asteroid surfaces, as well as other clues, scientists are able to estimate the depth of the regolith on various bodies. For Ida, Gaspra, and Eros, regolith depth is thought to be greater than 50 m in many places. For comparison, the lunar regolith is only 5–10 m thick. This is likely due to the relative infrequency of collisions on the Moon compared to what the asteroids experience, and the increased collisional speeds experienced by the Moon, which could serve to greatly increase the amount of material lost by the Moon rather than incorporated into the regolith.

CHARGED-UP DUST

High-energy ultraviolet light and x-rays from the Sun can cause particles in a regolith to have an electric charge. Because small bodies have so little gravity, the repulsion that same-charge particles have for one another in a regolith can be strong enough to lift regolith grains off of the surface entirely. This effect, called electrostatic levitation, has been observed on the Moon. While the theory is still being studied, levitation is expected to be strongest near the **terminator**, or the line on a body separating the day side from

Gold Dust

The first plans for landing on the Moon assumed the lunar surface would be bare rock. At the time, little was known for certain about the geologic history of the Moon, and the robotic *Surveyor* missions were designed to lead the way for Apollo. Among the critical questions about the Moon was whether there had ever been volcanic activity there, and whether the craters that dotted its surface were due to impacts or vulcanism. The prevailing view, which has turned out to be correct, was that the craters were due to impacts, but the low-lying dark lunar mare were in effect gigantic lava flows, and the lighter highlands areas were of a different composition.

Thomas Gold, an American astronomer, challenged this view. He argued that the lighter and darker areas were the same compositionally, but that meteorite impacts generated dust that moved downhill to the lower-lying areas, where the effects of solar wind and UV light darkened the dust. He argued that the mare could potentially have many meters of dust, with the consistency of quicksand, and that it would be shown to be unsafe to send astronauts. The *Surveyor* missions were built to deal with deep dusty surfaces as a result, though some skeptical scientists teasingly referred to the prediction as "Gold dust."

It was subsequently shown that the lunar surface is covered by a dusty regolith, but it has much greater packing strength than Gold feared, and the *Surveyor* and *Apollo* missions had no issues of sinking. While Gold was proven wrong in detail about the origin of the lunar mare and the depth of the regolith, he is credited with being the first to propose that a dusty surface layer might exist on planetary bodies.

night side. In theory, the very smallest particles on asteroids could be removed entirely, while particles larger than 1 µm or so can be moved from one part of an object to another, settling in lower areas or shadowed areas.

PONDS

Both electrostatic levitation and seismic shaking have been invoked to explain unexpected features on Eros and Itokawa called **ponds**. Although they don't contain water at all, their appearance gave them their name: flat expanses of fine-grained regolith that seem to fill depressions on the asteroid surfaces. Close-up images of ponds on Eros and Itokawa are shown in Figure 8.2. As expected from the previous discussion, the particles in the ponds on each asteroid are differently sized—Eros, with stronger gravity than Itokawa, also has smaller regolith particles. For comparison, Itokawa's regolith is composed of particles roughly the size of gravel, while on the Moon the regolith is more like talcum powder. On Eros, the particle size is intermediate between those cases.

Ponds on Eros can have sizes of 30 m diameter or more, with the largest ones found near the equator. Interestingly, the areas on Eros where ponds are formed tend to also be the areas with the lowest gravity and also ones

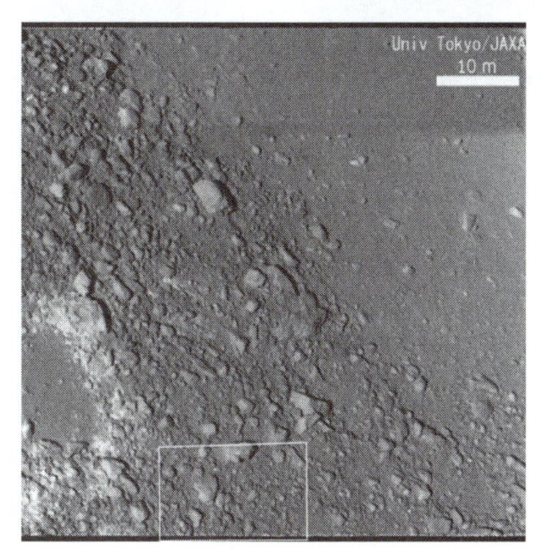

Figure 8.2 Smooth areas called ponds are found on Eros and Itokawa, the only asteroids observed at high spatial resolution. The top panel shows these ponds inside two craters on Eros, while the bottom shows the "Muses Sea," a smooth area on Itokawa, at the upper right. Also seen at the lower left on Itokawa is a bright-rimmed crater, one of the most obvious craters on Itokawa. NASA/JHUAPL.

that spend a large fraction of Eros' year near the terminator due to Eros's unusual orientation, with its pole nearly in the plane of its orbit. These factors seem to support an origin via electrostatic levitation. Itokawa is much smaller than Eros, and only has two areas that could be considered ponds. In contrast to Eros, these areas are near the poles on Itokawa. While analysis of ponds on Eros and Itokawa is still ongoing, it seems likely that additional data from more asteroids will be necessary to figure out how they formed.

GROOVY SURFACES

An unexpected finding from early images of Mars's satellite Phobos were straight-line segments that in some cases seemed to radiate from Phobos's

Figure 8.3 Phobos, the inner satellite of Mars, has an extensive set of grooves on its surface, visible here stretching horizontally. Grooves are also seen on the asteroids Ida, Gaspra, and Eros. Their origin is not completely solved, but the leading theory is that they represent sub-surface fractures into which regolith has partially drained. NASA/JPL/Malin Space Science Systems.

largest crater. These segments, known as **grooves**, have been a matter of some controversy. In addition to appearing on Phobos, grooves are also seen on Ida, Gaspra, and Eros. On Phobos, grooves have been attributed to many processes including explosive sublimation of internal ice (somewhat similar in concept to some of the cometary and dwarf planet processes described later) and resulting from fast-moving impact ejecta from craters on Mars. Their presence on Gaspra, Ida, and Eros, which are not thought to have internal ice and which obviously are not orbiting a planet, require a different explanation. On Gaspra and Ida, some of the grooves appear as strings of pits roughly 100 m wide and up to a few kilometers long. The depths on Ida seem to be a few tens of meters, though that information isn't available for Gaspra.

As discussed further in Chapter 9, many asteroids are thought to have deep fractures that can cut through the entire object. This has led to an alternate explanation for grooves—as areas where fractures are present beneath the regolith. In this interpretation, regolith drains partway down the fractures, resulting in the grooves. Models suggest that the appearance of a row of pits also fits this explanation. More recently, the Rosetta flyby of the asteroid Steins showed an object with a line of craters that seems to radiate from a large central crater. While analysis is still underway, this too is consistent with the overall picture. Itokawa seems to lack grooves, consistent with the idea that rather than consisting of relatively intact pieces separated by fractures, it is an unconsolidated rubble pile composed of smaller re-accumulated fragments with no overall structure.

MASS WASTING

The general term for material moving downhill on the surface of a body due to gravity is **mass wasting**. This is a common occurrence on the Earth, and there is evidence it occurs on small bodies as well. Because asteroids can have regoliths of loose material, we can imagine their very surfaces as acting like exceedingly dry sandboxes. Any of us who have spent any time in a sandbox know it is impossible to create sheer cliffs in dry sand. We can build a pile of sand, but the height of that pile is related to the size of its base. Once the slope of the pile becomes too steep, sand will cascade down the sides of the pile until the slope is restored to a critical angle or less. This angle is called the **angle of repose**, and it is important on small body surfaces.

If we consider the physics of the angle of repose, we find that it is, perhaps surprisingly, independent of the strength of an object's gravity. Therefore, it is relatively straightforward to calculate. It is, however, not independent of strength. A pile of sand, which has no internal strength, is very different from the solid rock that makes up mountains and canyons. Spacecraft looking at the surfaces of asteroids and comets have enabled maps to be made of the distribution of slopes. Almost always, they find the slopes are less than the angle of repose, which is consistent with the idea that their surfaces are covered with loose regolith rather than solid rock. An intriguing exception is Comet Wild 2, found by *Stardust* to have some areas that must be made of strong material, presumably a combination of rock and ice.

MATURITY AND IMMATURITY

To understand the regoliths of the small bodies, we must first turn to the Moon. One of the unexpected findings from the Apollo missions was that regolith had some properties that were very different from the rocks from which they were derived. The spectral properties of regolith and rocks were similar in general, but different in detail, appearing darker and redder than the original rocks, as further discussed in Chapter 7. Geochemical analyses showed that some regolith had accumulated relatively large amounts of hydrogen. Further research has showed that these changes are due to exposure of the regolith to micrometeorite impacts and the solar wind. The solar wind is composed mostly of hydrogen, which sticks more easily to the grains in regolith than to rocks as a whole, explaining the increased amount of hydrogen in regolith. The spectral changes appear to be due to micrometeorite impacts, which vaporize a tiny amount of material and create iron vapor, which can then solidify and coat regolith grains, altering their spectra. In the lunar regolith, samples which have more solar wind implanted hydrogen and redder and darker spectra are called more **mature**, while those that are closer to the initial rocks in their properties are called **immature**. Eventually, undisturbed regolith on the surface of the Moon will be unable to hold any more hydrogen and will be fully coated in iron. This process can take 1 billion years or longer.

However, lunar regolith is rarely completely undisturbed. The processes that mature the lunar soil only act on its very surface, in the top few thousandths of a centimeter. Below that top layer, material is largely protected from the solar wind and micrometeorites. Larger impacts, however, will easily penetrate much deeper. An impactor only 10 cm in diameter (roughly hand-sized) will result in a crater that would be waist-deep for most adults, and the formerly protected, immature material inside the crater will find itself on the surface. This process, called **gardening**, is constantly occurring, mixing more-mature regolith with immature regolith.

The same processes that are occurring on the lunar surface are thought to occur on airless bodies throughout the solar system. As a whole, the processes that affect and mature small body regoliths are known somewhat confusingly as space weathering. Because there is no weather in space, this name can be seen as an unfortunate one!

Because we do not have samples directly from asteroid, comet, or dwarf planet surfaces, our ideas of space weathering on small bodies are still uncertain. There has been circumstantial evidence for decades that the spectra of asteroids do not match the meteorites we believe they provide, at least in some cases. Spacecraft visits to Gaspra and Ida, and notably Eros and Itokawa, showed scenes consistent with space weathering—dark regolith sliding downhill revealing bright regolith, and fresh-looking areas being brighter and less red than less-fresh areas.

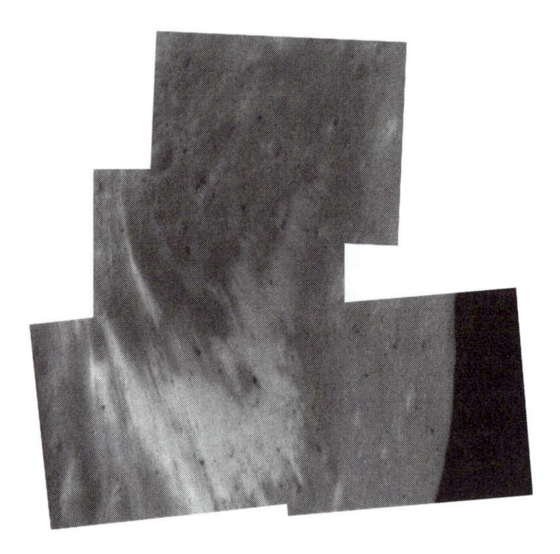

Figure 8.4 This image of Eros' surface from the *NEAR Shoemaker* spacecraft shows a view across a crater. Bright and dark areas are shown, with darker spots at the bottom of the crater and outside the crater. This is interpreted as due to originally brighter material darkening with time via "space weathering," and then slumping to the bottom of the crater, leaving bright material visible on the walls of the crater. NASA/PDSSBN.

The extent and importance of space weathering on asteroids has been a matter of controversy for decades. It has been argued that because the Moon is much closer to the Sun than the asteroid belt is, and the speeds with which micrometeorite collisions occur is much faster at the Moon, that space weathering shouldn't be as effective. An additional unknown factor is the different compositions of the Moon and asteroids, which is thought to result in changes in how their regoliths might mature. Evidence for space weathering is largely absent from the meteorites. Some meteorites, called regolith breccias, have spent time at the surface of an asteroid, and show evidence for increased amounts of hydrogen and other gases from the solar nebula, but show no evidence for vapor-deposited iron. On the other hand, it has been suggested that the processes that turn the powdery regolith into a rock strong enough to survive the passage to Earth and become a meteorite might also destroy much of the evidence of space weathering.

Circumstantial evidence for space weathering on asteroids using ground-based telescopes has also been found. It has been observed that on average, small near-Earth asteroids (NEAs) are less red and have higher albedos than larger ones. Because small objects are broken up and destroyed by collisions more frequently than larger ones, scientists can use the sizes of NEAs to guess their relative ages. (It should be stressed that the meaning of "age" for small bodies depends on the context. The minerals found in all small bodies of all sizes formed billions of years ago. When discussing the age of an asteroid surface, we usually are describing the time since it has been exposed to space. In the case of large and small asteroids, we are describing the time since they have reached their present sizes.) When putting all of this information together, the data are consistent with a scenario where younger objects have less mature regoliths that become more mature with time, consistent with what is seen on the Moon. Critics note, however, that this is hardly the only way to explain the data, and that the space weathering interpretation is based on a number of assumptions that may or may not be true.

In an effort to better understand what is going on, scientists have attempted to re-create space weathering in their laboratories. The most promising experiments have involved firing a laser at a simulated regolith for very short periods of time—milliseconds or less. These very short bursts are used to try and replicate the tiny amounts of heating and melting experienced in micrometeorite impacts. The results are comparable to what is expected of space weathering of asteroids: reddened and darker materials.

Because the results of space weathering are most obvious for ordinary-chondrite-like compositions, most experiments have been on those meteorites and the minerals in them (like olivine and pyroxene). However, micrometeorites and the solar wind should also be affecting the surfaces and regolith of all asteroids, regardless of composition. Scientists are only now beginning to perform space weathering experiments on different meteorite types, including the carbonaceous chondrites. Interestingly, there are hints that the spectral effects of space weathering may be very different for these objects, and that the destruction of organic material in outer belt asteroid regolith may make these objects *less* red, in contrast to the inner-belt asteroids.

Icier objects in the outer solar system can also be affected by the ultraviolet light of the Sun and cosmic rays originating from outside the solar system, in addition to micrometeorites and the solar wind. With long times between collisions, astronomers expect that objects like the transneptunian objects (TNOs) experience very little gardening. Experiments to simulate space weathering with the theorized composition of TNOs and the conditions they experience show that their spectra darken and become less red with increased exposure.

COMETARY SURFACE PROCESSES

Cometary surfaces are, in contrast to asteroidal surfaces, in a constant state of flux. They suffer impacts and accumulate craters, but they also experience erosion and internal modification. The surface of a comet is intimately connected to its coma, serving as the coma's area of origin.

When far from the Sun, the surface and near-surface of a comet is cold enough that any ices present are stable as solids. At these temperatures, their surfaces experience the same sorts of processes as asteroids, though space weathering is expected to be a very slow process at great distances from the Sun. Nearer the Sun, however, temperatures rise high enough that ices are no longer stable. In the vacuum of space and the low pressures present just below cometary surfaces, liquids are also unstable and solids sublime, or change directly into gas. This gas escapes, carrying some dust and minerals with it and creating the coma.

It is probably obvious that the escape of gas and dust from cometary surfaces will also help to erase craters and other surface features. However, the loss of ice from comets does not occur evenly across their surfaces. It has been known for decades from telescopic observations that much of the gas and dust comes from a small number of relatively compact areas. These **cometary jets** cover only a small fraction of the comet's surface area, as little as 5 percent in some cases. The rest of the surface can be relatively inactive.

The rate at which dust is carried off with gas into the coma is quite important. If the subliming ice carries a lot of material with it from an active area, there will be a constant supply of fresh ice seeing sunlight for the first time. If, on the other hand, the gas mostly escapes free of dust, the ice level will retreat further and further beneath the surface. As a result, the near-surface regolith of a comet can become ice-free. Such an ice-free regolith, called a **lag deposit**, is an effective insulator. As the comet approaches the Sun, solar heating is unable to penetrate beyond this lag deposit to the fresh ice below, or penetrates slowly enough that the ice remains frozen until the comet recedes from the Sun and temperatures abate.

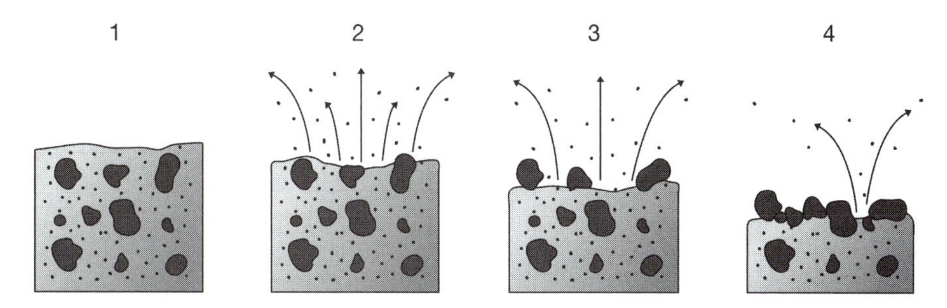

Figure 8.5 This series shows the creation of a lag deposit on a comet. A cross-section of the original cometary surface is shown at left, with a mixture of ice, dust, and larger rocks. As the comet approaches the Sun, ice sublimes, taking some dust with it (panels moving to the right). The larger rocks, however, cannot be moved by the escaping gas, and collect at the surface. Eventually, the rocks begin to block the ice sublimation, and the gas production ceases from that area of the comet. The layer of rocks is called a lag deposit. If gas production ceases on the entire cometary surface, the object is considered an extinct comet. Illustration by Jeff Dixon.

Comets from the transneptunian region or the Oort cloud making their first visits to the inner solar system will have a large amount of ices near (or at) their surfaces, with no lag deposit at all, and can be extremely active. This was the case with the recent comets Hale-Bopp, Hyakutake, and McNaught. With repeated passes by the Sun, the lag deposits grow thicker. Comets that have been perturbed by close passes to planets in the inner solar system or near-Earth orbits will experience higher temperatures for longer periods of time, and may reach the point that the lag deposits are thick enough over the entire surface to prevent any activity at all. As touched on in Chapter 1, these objects are compositionally still comets, but are no longer observationally any different from asteroids. Such bodies are called **extinct comets** and are thought to represent somewhere between 5 and 20 percent of the near-Earth object population.

Spacecraft images of comets show surfaces that are superficially like what is seen on asteroids, but with obvious differences. Comet Wild 2, seen by the *Stardust* mission, has areas that look crater-like, but have depths that are very unusual for their sizes compared to craters on asteroids, or other planets and satellites. These are thought to be craters that have been modified by erosion of the surface due to gas and dust loss. Comet Borrelly, visited by *Deep Space 1*, found a largely smooth surface devoid of craters but with some evidence of low cliffs from which jets were originating, as well as hints of layering. The Deep Impact images of Comet Tempel 1 had characteristics of Borrelly and Wild 2, with some smooth surfaces, layers, and cliffs of roughly 20 m, as well as some craters on other parts of its surface.

GAS-POWERED MOTION

The mass loss that comets experience means that they are constantly, if slowly, changing shape. These shape changes slowly change the rotational properties of a comet, bringing increased sunlight to some areas, and decreased sunlight to others. The mass loss itself also is a factor independent of the shape changes. Because the dust and gas carries momentum, the comet's orbit slowly changes, with greater changes coming with greater activity. These **non-gravitational forces** are hard to predict, but must be accounted for in precise calculations of cometary orbits.

In addition, the amount of gas loss places a rough lifetime on comets. For instance, it is estimated that Comet Tempel 1 loses roughly 1 billion kilograms of mass during each orbit, about the mass of the Great Pyramid of Giza in Egypt. The mass of the entire comet is only 10,000 times greater, meaning that after roughly 10,000 orbits the mass in the comet would be entirely ejected. With an orbital period of less than six years, we might expect Tempel 1 to be absent from the sky within roughly 60,000 years. Long before then, however, Tempel 1 will likely break into smaller bits rather than steadily lose mass to all directions, and those smaller bits will likely evaporate more quickly. Alternately, as already described, Tempel 1 could become an extinct comet.

DWARF PLANET SURFACE PROCESSES

As larger objects, dwarf planets can have some surface processes similar to those seen on full-fledged planets or the largest satellites, in addition to many of the processes seen on the smaller asteroids and comets. However, none of the dwarf planets have been visited by spacecraft, leading scientists to make their best guesses by studying objects thought to be similar.

For Pluto, this means a comparison to Neptune's satellite Triton. Triton and Pluto are roughly the same size and at similar distances from the Sun. Both are thought to have similar compositions. *Voyager 2* passed Triton in 1989 and returned high-resolution images of its surface. The results were surprising, as evidence of **cryovulcanism**—volcanoes that erupt water or methane rather than rock—was seen on Triton's surface. Just as volcanoes on the Earth change its surface, cryovolcanoes on Triton serve to change its surface by erasing craters and building mountains.

It is not clear where the energy for Triton's volcanoes originates. On the Earth, the heat is leftover from long-lived **radionuclides** (further discussed in Chapter 9). On active satellites like Io and Enceladus, volcanic activity is thought to be powered by tidal forces and the gravitational pulls of their central planets and other satellites. On Triton, however, the idea with the most support is the **solid state greenhouse effect**. We are familiar with the greenhouse effect, which is thought to be a major reason for climate change on Earth: gases such as carbon dioxide and water vapor allow visible light to pass through and heat the ground, but are efficient at trapping reradiated heat, which leads to an increase in temperature.

The concept of the solid state greenhouse effect is similar, except that instead of gases in an atmosphere, it is solid nitrogen in the ground itself that allows visible light through and traps the heat. Eventually, areas below the surface heat up to the melting or boiling point of nitrogen and erupt.

Without close-up images for Pluto, Eris, or other transneptunian dwarf planets, we are left to hypothesize based on Triton. Observations of crystalline water ice on Pluto's satellite Charon have been interpreted as requiring cryovulcanism, though there may be other explanations. Searching for evidence of cryovulcanism on Pluto is one of the goals of the *New Horizons* mission.

We do not have close-up spacecraft images of Ceres, nor do we have any good ideas of what objects might be similar. There are some Hubble Space Telescope views of Ceres' surface, which have tantalizing clues, however. There are circular areas that appear consistent with craters, and some brighter and darker areas in general. However, the difference in brightness across Ceres' surface is quite subtle.

In anticipation of Dawn's arrival at Ceres, scientists have attempted to predict what Ceres' surface will be like. Compositional studies like those

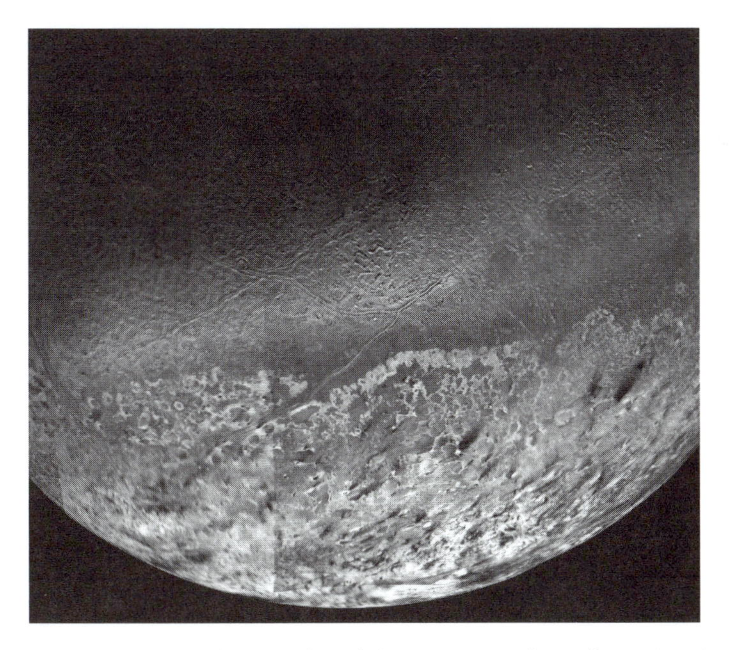

Figure 8.6 Although not classified as a dwarf planet, Neptune's satellite Triton is thought to be similar to Pluto. This image of Triton's surface, from *Voyager 2*, shows a large number of dark streaks, thought to be clouds from erupting volcanoes blowing in Triton's thin atmosphere. These volcanoes are thought to be due to the heating and sublimation of subsurface nitrogen ice, which eventually builds up enough pressure to erupt. NASA/JPL.

described in Chapter 7 show that despite an ice-rich interior, Ceres does not have an icy surface, and calculations of Ceres' surface temperature show that ice is only stable very close to its poles. However, there is a chance that frost can form on shaded slopes. There is also an interesting suggestion that Ceres' surface may be covered in cracks. We know that ice is less dense than water (we see it every time we put ice cubes in a punch bowl or glass of lemonade). This means that for a given mass of water, its volume increases as it freezes. Observations of Ceres suggest that it has a deep layer of ice, discussed further in Chapter 9. At one time that layer would have been liquid, meaning that as it has frozen its volume would have expanded. This would have led to stresses at Ceres' surface, perhaps resulting in huge faults and cracks. An additional factor is the instability of ice on Ceres' surface, as mentioned before, which might lead to some of the same processes on Ceres' surface as are seen on comets—a lag deposit, for example.

SUMMARY

The surfaces of the asteroids, comets, and dwarf planets have been affected by both internal and external processes. All have suffered impacts, with the larger objects thought to have accumulated a blanket of broken-up material, or regolith. Craters appear on all of the small bodies for which we have close-up images. However, sublimation of ices has modified cometary surfaces and erased craters on some of their surfaces. The regolith on small bodies has redistributed itself due to shaking during impacts and simply moving downhill, as well as perhaps via more unexpected forces like electromagnetic levitation. While dwarf planets are largely unexplored, conjecture based on similar objects suggests the possibility of volcanoes and more complex surface topography.

WEB SITES

A technical description of seismic shaking on asteroids, including a set of explanatory images from *NEAR Shoemaker* and other missions: http://www.astro.cornell.edu/~richardson/asterseismo.html.

Professor Dave Jewitt of the University of Hawaii provides information about comets and dwarf planets at this Web site, including a simplified explanation of how lag deposits form on comets: http://www.ifa.hawaii.edu/faculty/jewitt/rubble.html.

9

Small Body Interiors

While the surfaces of objects are visible to our telescopes and spacecraft, the interiors of small bodies are of intense interest to scientists as well. Learning about a body's interior gives us information about the amount of heating it has experienced and how consolidated it is. The way an object reacts to an impact is critically dependent on its interior configuration. It is also of critical concern for those studying the best way to defend Earth against NEO collisions. Through a variety of direct and indirect measures, we have learned a great deal about what lies beneath the surface of asteroids, comets, and dwarf planets, although there is a great deal that is still uncertain. In this chapter, we look at the interior structure of small solar system bodies and what that tells us about their formation and nature.

STRENGTH, POROSITY, AND RUBBLE PILES

The interiors of objects are critically affected by the **strength** of the rocks and minerals that are present. The strength of an object is most simply described here as the amount of force per area of material that can be applied before it breaks or deforms. There is more than one kind of strength, depending on the direction from which the force is applied. The two that concern us here are compressive strength, the ability to resist crushing, and tensile strength, the ability to resist being pulled apart.

The strength of a planet-sized object increases with increasing size. This is because with increased size comes increased gravity, which holds planets together. The theoretical force required to "break" a planet is exceedingly large, but is larger for Earth than it is for Mars, and larger for Mars than it is for Mercury. Similarly, it is larger for an object the size of Ceres (nearly 1,000 km in diameter) than it is for something only 100 km across. At the distances in the Kuiper belt and through the transneptunian region, water ice is so cold it behaves like rock, so the same statement holds for those bodies as well. If we extrapolate this behavior to smaller and smaller sizes, we expect small objects to be much *weaker* than large ones.

We can directly measure the strength of small objects in the laboratory. For those samples, the size of one's hand or smaller, strength is observed to *increase* as they get smaller. This somewhat surprising result is because the mineral crystals that make up rocks can be quite hard to crush or crack, but the collections of minerals are much easier to separate from one another. The presence of small cracks in rocks reduces their strength, and as rocks get to be the size of meters or larger, they have more cracks and a greater chance of a force applied from outside randomly aligning with some of those cracks. If we extrapolate this behavior to larger and larger sizes, we expect small objects to be much *stronger* than large ones.

Obviously, small objects cannot be both weaker and stronger than large objects, though that's what the data suggest. The problem lies in extrapolating too far. Scientists believe that as size increases from centimeter-sized objects, strength decreases until it reaches a minimum at some critical size, beyond which strength again begins to increase. The two regions on either side of the critical size have different names: the side with the largest objects is called the "gravity regime" since their larger gravity is important; the side with the smaller objects is called the "strength regime." The value of the transition size is still a matter of debate for scientists, but is thought to be somewhere in the range of 100 m.

Usually scientists are most concerned with compressive strength, which is most relevant to impacts. However, tensile strength is what holds objects together when they are experiencing tidal forces while passing close to a planet or the Sun. Models of cometary breakups under those conditions suggest that comets have exceedingly low tensile strengths, much lower than even snow. These results are complicated by the fact that comets may be poorly held together to begin with (see the following discussion), but the conclusion is certainly that comets are not strong objects by any stretch of the imagination.

As mentioned before, even solid rock has spaces and tiny cracks. This empty space is called **porosity**. A pile of rocks has porosity as well, also called **macroporosity** since it is between rocks rather than within a rock. Such a pile typically can have a porosity of 40 percent or more, so 40 percent of the volume is empty. A pile of sand will have a much lower porosity. Two objects of the same composition but different porosities will have different densities.

There are relatively few asteroids or comets for which we have measured densities and known compositions. For those bodies, we have good

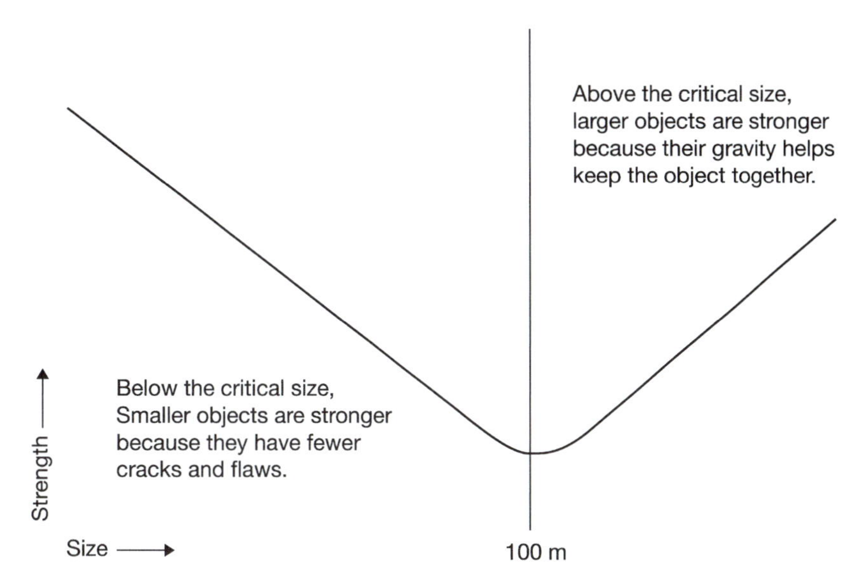

Figure 9.1 Laboratory experiments on rocks and minerals show that their strength decreases as they get larger, due to the larger number of pre-existing cracks and flaws. On the other hand, theoretical calculations and consideration of gravity suggests that larger planets will be stronger than smaller ones because the force of gravity on larger objects will make them harder to "break" and disperse. These two facts have led to predictions that stony objects of roughly 100 m in diameter will be the easiest to disperse in terms of energy needed per gram of material. Larger objects are considered to be in the so-called gravity regime, smaller ones in the strength regime. Illustration by Jeff Dixon.

estimates of porosity. There are a greater number of objects, however, for which we have estimated compositions that can give us estimated porosity. We find that the largest, most massive objects have low porosities. This is not surprising, since their gravity is large enough to crush empty areas, much the same way pushing on a sleeping bag or air mattress will make them denser. On the Earth, near-surface porosity can be high since there isn't much pushing on any void space from above. Deeper in the ground, the mass of rock pushing from above increases until void spaces cannot support the weight above. As gravity becomes weaker on smaller bodies, the weight of overlying rock becomes less, and void spaces can exist further below the surface, until at last objects are small enough that voids can be maintained through the entire object.

This can be thought of as similar to the way a house might react in a similar situation. A house is mostly empty space that is maintained by the floor, walls, and roof. If enough mass is piled on top of the house, eventually the roof will give way and crush the rooms inside. This is also true if gravity were to be magically increased—eventually, the force from the roof itself would be too much to be supported and the house would crush itself.

At the very smallest sizes, we expect objects with diameters of only a few meters to tens of meters (or smaller still) to have low porosity, since they are almost certainly single rocks much like similar-sized rocks on the Earth. Such objects are called **monoliths** and are thought to have both compressive and tensile strength. If a monolith were to be thoroughly cracked and fractured, its tensile strength would disappear, but it still would have some compressive strength. Such a body is called **shattered**. Finally, if a shattered body were to be taken apart, for instance via impact, and reassembled differently, with pieces moved and rotated relative to their original positions, the resulting body would react very differently to stresses than a monolith or shattered body, absorbing and localizing stress more than transmitting it. These bodies are called rubble piles, and have significant macroporosity, as described previously.

DIFFERENTIATION, HEATING, AND COOLING

As discussed in Chapter 4, the small bodies accreted from pieces that were of uniform composition depending on where they formed in the solar system. If we look at a chondrite in detail, we find it composed of minerals of very different densities, from grains of iron-nickel metal to crystals of olivine. If we were to look at a body like Itokawa, or a parent body of chondritic meteorites, we would find the same composition throughout, regardless of depth. In the outer solar system, temperatures would be cold enough to have accreted ice and other volatiles, and those would be found mixed in with silicates and metal as well. This increases the range of mineral densities found in close proximity even more. An object that has such a mixture of minerals, and which has the same composition throughout, is called **undifferentiated**.

All solar system bodies accreted some amount of radioactive materials. As those materials decay, they give off heat. This **radiogenic heat** will tend to heat up the body. All objects in the solar system began with roughly the same concentration of radioactive elements, whether ^{26}Al, which has almost entirely decayed at this time, or ^{238}U, which today is still decaying and giving off heat. Because this concentration is roughly the same, it will increase at the same rate as an object's mass—a body 10 percent the mass of the Earth will have accreted 10 percent of the Earth's amount of radioactive elements, and a body 10 times larger will have accreted 10 times the amount of Earth's radioactive elements.

Objects in the solar system lose heat by radiating infrared light into space. The amount of energy radiated increases as the surface area of a body increases—larger areas mean more room is exposed to space, which allows heat to escape. How do these two effects balance out? As noted, radiogenic heating increases with the mass of an object:

1. Heating α Mass

The density of an object is a well-known quantity,

2. Density = Mass/Volume

and the volume of a sphere is also a well-known quantity:

3. Volume = (4/3) $\times$ π $\times$ (radius)^3

Using these equations we can relate how the radius of an object changes as its mass increases:

4. Mass α Density $\times$ Volume = Density $\times$ (4/3) $\times$ π $\times$ (radius)^3

For our purposes, we can ignore the constant terms and be satisfied by saying:

5. Mass α Density $\times$ (radius)^3

So

6. Heating α Mass α radius^3

Increasing the radius of an object by a factor of 10 will increase the heating by about a factor of 10^3, or 1,000.

What about cooling? Again as discussed here, as the surface area increases, so does the energy radiated:

7. Cooling α Surface Area

Once more, the surface area of a sphere is known:

8. Surface Area α 4 $\times$ π $\times$ (radius)^2.

So we can quickly determine, again discarding the constant terms, that

9. Cooling α Surface Area α radius^2.

Thus, increasing the radius of an object by a factor of 10 will allow cooling to occur 100 times faster. But we already know that increasing by that size will lead to 1,000 times more heating! So as objects get larger, they tend to retain their heat more efficiently. The smallest objects can cool very rapidly and never get very hot to begin with, while objects the size of Earth will warm up from the decay of their radioactive materials. While the rate of heating from the long-lived radionuclides still decaying today is not sufficient to heat dwarf planets and large asteroids, short-lived radionuclides gave off a tremendous amount of heat early in solar system history, which could be retained by bodies the size of Vesta, Pluto, or Ceres. Typically, temperatures would have increased as the short-lived nuclides gave off heat faster than it could be radiated until some maximum temperature was reached. As time went on, there was less remaining radioactive material, so the rate of heating went down. Eventually, objects were able to radiate heat faster than they were generating it, and they began to cool. The maximum temperature that was reached is dependent upon the size of the object, the length of time it took to accrete, and other factors that are more difficult to determine like the amount of regolith on the surface, which tends to insulate objects compared to those objects without regolith. Some of the short-lived radionuclides have half-lives so short that we expect none of the atoms to remain. For instance, ^{182}Hf has a half-life of 9 million years. After 500 half-lives, the amount remaining is too small to be detectable. These so-called extinct radionuclides can still be studied because of the excess of the stable decay products that they left behind, in this case ^{182}W.

As noted before, objects in the asteroid belt formed as mixtures of rock and metal, some of which survive as the parent bodies of the chondritic meteorites discussed in Chapter 4. Smaller objects reached maximum temperatures that were relatively cool. At some critical size, however, the maximum temperature became hot enough to melt first the metal and then the rock. When this happened, the liquid metal rapidly began to sink through the liquid rock since it is so much denser. As a result, these objects were left with a metal core and a rocky exterior, or **mantle**. This molten mantle often would experience further volcanic processes and separate further, with a low-density **crust** atop an olivine-rich mantle. This process is known as differentiation, and objects that have experienced it are fittingly called differentiated.

Objects that formed where ice is stable could experience an additional differentiation before the rock-metal separation described previously. When the melting temperature of ice was reached, a layer of liquid water could form. The melting temperature of ice is, as we know, much lower than the melting temperature of metal, so the rock and metal remain mixed with each other, at least at first. In this case, the remaining rock-metal mixture would sink, leaving the water as the highest layer. Depending on the duration of the heating, the amount of initial water ice, and the size of the object, the end result could be an icy mantle over a rocky center, a full separation of ice, rock, and metal, or the loss of some or all of the water and a rock-metal object like those described before.

ROCK/METAL AND ICE/ROCK RATIOS

Density is the best single measurement with which to understand the interior structure of an object. Armed with just the density of an object and a few guesses about its composition, we can come up with a pretty good, if rough, idea of some of its interior properties, and knowing the relative densities of materials lets us know which materials are closer to the surface or deeper in an object. As an example, let's look at Vesta. We suspect for a number of reasons explained elsewhere that Vesta is a differentiated object, with an iron core and rocky mantle and crust. About how big is its core and how much of Vesta's mass is it?

To begin, we consider Vesta's density (about 3,400 kg/m^3) and the typical densities of iron metal and rock (8,000 and 3,000 kg/m^3, respectively). Because Vesta is so large, we'll be able to ignore its porosity, something we would have to account for in some smaller bodies. Since we're considering only rock and metal, Vesta's bulk density will depend only on the relative proportion of these two components, and the fraction of rock will equal 1 – the fraction of metal.

10. Volume fraction of rock + volume fraction of metal = 1.
11. Density of Vesta = density of rock × volume fraction of rock + density of metal × volume fraction of metal.
12. Density of Vesta = density of rock × (1 – volume fraction of metal) + density of metal × volume fraction of metal.

Since we know most of the numbers we need, we can pretty quickly solve for the fraction of metal in Vesta:

13. 3,400 = 3,000 × (1 – volume fraction of metal) + 8,000 × volume fraction of metal.

The fraction of metal works out to roughly 0.08, so 8 percent of Vesta's volume is found in the metal core. However, we're interested in the size of the core, so we're not quite finished. Because volume goes as radius cubed (volume αr^3), we take the cube root of 0.08 to find the fractional radius of Vesta's core and find it to be 0.43. With Vesta's radius of about 265 km, the radius of the core is thus roughly 115 km.

What about the mass fraction? With the volume fraction of the core and knowing that density = mass divided by volume, we can find that the

14. Mass of core = Density of core × Volume of core
15. Mass of Vesta = Density of Vesta × Volume of Vesta
16. Mass of core/Mass of Vesta = (Density of core × Volume of core) / (Density of Vesta × Volume of Vesta) = (Density of core / Density of Vesta) × Volume fraction of core
17. Mass of core/Mass of Vesta = 8000/3400 × 0.08 = 0.19.

So roughly 20 percent of Vesta's mass is in its core. While we liberally rounded off numbers in the previous equations, and even though we ignored the fact that Vesta isn't *quite* spherical, the answer we determined is within the range of core sizes scientists expect for Vesta.

A similar exercise can be performed on icy bodies, substituting the density and fraction of ice for the density and fraction of metal, and remembering the rock is in the center rather than on the outside in that case! More complicated combinations of components can also be performed, for instance for a body with ice, rock, and metal, though the results become more uncertain as additional ingredients are added. As mentioned already, porosity needs to be considered for smaller objects, and as bodies become planet-sized, the density of rock can change as depths increase, which makes the calculations more complicated. As a quick way of understanding objects, however, this simple calculation is a good tool.

VESTA IN DEPTH

Of course, scientists have used more complex techniques and have learned a great deal more about Vesta than simply an estimate of its core size. We are immensely helped by the existence of the HED meteorites and the recognition that they come from Vesta. Among the data used to deduce that Vesta has a core is the fact that the HED meteorites are the products of melting and they are missing siderophile elements, or those that chemically are more likely to be found with metal, relative to the amount of those elements found in the undifferentiated, chondritic meteorites. This means that a metallic core was formed, taking the siderophile elements as it formed. Radioactive elements can also be siderophiles or lithophiles, and in some cases they decay from one type of element into another. This allows further insight into the core formation process. For instance, Hf is a radioactive lithophile element, while W is a siderophile. At the time of core formation on an object, the W would be taken into the core, while Hf would stay behind in the crust and mantle. After core formation, radiogenic ^{182}W would continue to be created from ^{182}Hf, however. By studying the amount of ^{182}W found in the HED meteorites, geochemists have found that the differentiation of Vesta occurred very early on, within the first 5–15 million years of solar system history.

Scientists suspect more than 100 objects differentiated in the asteroid belt, given the characteristics of the iron meteorites that have been analyzed. These objects are difficult to find in space, however. We might expect that with 100 differentiated objects, there should be 100 objects like Vesta present in the asteroid belt. However, Vesta is unique in being a very large object nearly unanimously considered to be differentiated. One implication is that the other differentiated objects may have been victims of catastrophic collisional disruption, with only fragments remaining today. Remote sensing studies of the type described in Chapter 7 have been used to try and identify

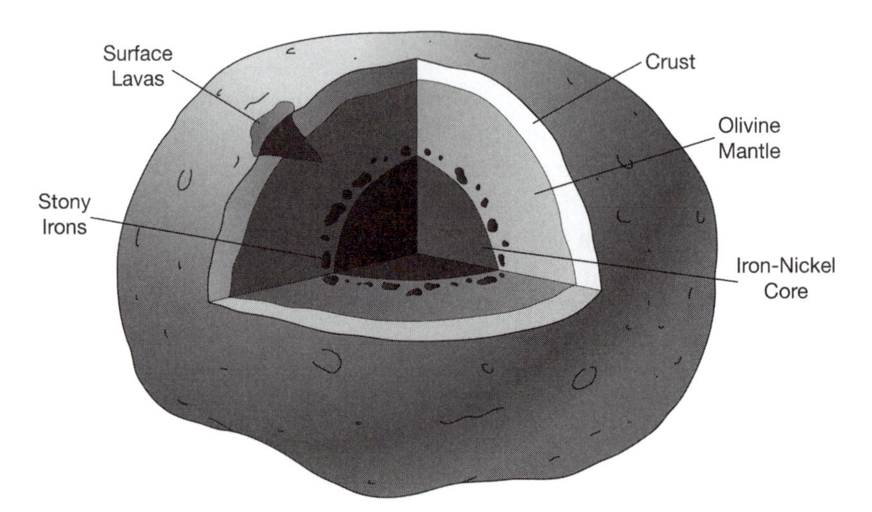

Figure 9.2 This cartoon shows our current model of Vesta's interior. From our knowledge of meteorites, Vesta's composition, mass and density, we can estimate its metal-silicate ratio. Even without the kind of data available for the Earth and other planets, we are fairly certain that Vesta has an iron-nickel core, an olivine mantle, and a basaltic crust. Illustration by Jeff Dixon.

metallic objects, which would mostly likely be the cores of disrupted bodies, and allow us to study a planetary core in detail.

CERES

While Vesta is dry, Ceres has long been known to contain water in some form. Using similar calculations as done in the example for Vesta, we find Ceres to have 15–20 percent water by mass. As mentioned in Chapter 6, the shape of Ceres has been determined using images from the HST. The shape of an object and its moment of inertia are related in an intimate way to its interior structure, with different core densities and radii resulting in different observable shapes.

Using the HST data, scientists found that Ceres has a rocky core with an icy mantle 65–125 km thick not far below the surface. This is consistent with computer models of the thermal evolution of Ceres. These models start by assuming Ceres originally had certain fractions of rock and ice, and certain concentrations of radioactive elements, and then calculate how its interior would change as those radioactive elements decay and give off heat. One of the critical unknown quantities is the amount of ammonia (NH_3) that is present in Ceres. Because ammonia is a "natural antifreeze," its presence makes it more difficult for water to freeze, and if enough is present on Ceres, it is possible some liquid water still exists in its interior. The thermal

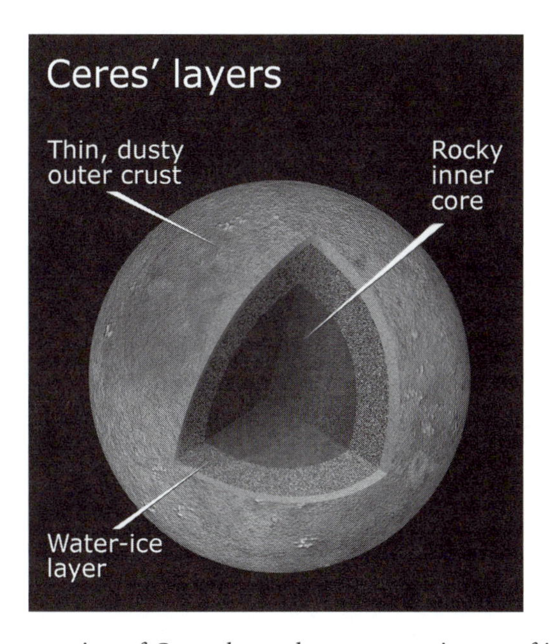

Figure 9.3 This cutaway view of Ceres shows the current estimates of its interior structure: A rocky core covered with a water ice mantle, and finally a thin outer crust. These estimates are made by knowing Ceres' density and the density of the likely constituents, as well as measures of Ceres' shape from the Hubble Space Telescope. STScI.

models also suggest that Ceres could possibly be differentiated further, with an iron core of roughly 100-km radius, similar to what is expected for Vesta. Current research is investigating whether some other large asteroids like Pallas and Hygiea could also have similar icy mantles.

THE TRANSNEPTUNIAN DWARF PLANETS

Although they are largely unknown in many ways, scientists can make certain educated guesses about the interiors of Pluto, Eris, and other large TNOs. As with Ceres, estimates of the ice/rock ratio for these objects can be made (roughly 30–50 percent ice on Pluto, 35–80 percent on Eris). Given the amount of radioactive materials that these objects are expected to have accreted and their size, both Pluto and Eris, as well as other large TNOs, are expected to have differentiated their ice from their rock, though not to have gotten hot enough to have separated a metal core. Our knowledge of the TNOs is limited by their distance (the large spread of possible ice/rock ratios for Eris is because we have been unable to measure its volume as precisely as other objects) and also our limited knowledge of internal compositions. Similarly to Ceres, if Pluto or Eris accreted enough ammonia, they may still have subsurface oceans of liquid water. Other materials expected to be present in the larger TNOs, like methanol (CH_3OH), will also lower the freezing point of water.

The large TNO Haumea has a surprisingly large density of 2,600 kg/m^3, implying a very low ice/rock ratio of only 15 percent or so. Haumea has two satellites and several objects sharing similar orbits in a dynamical family, believed to have formed via a large impact early in solar system history. Because its density is so high, scientists have concluded that Haumea was differentiated at the time of the impact, which is thought to have stripped away much of its icy mantle, resulting in the low ice fraction we see today.

As with the asteroids, smaller TNOs cannot retain heat effectively, and are thought to be more or less as they were early in solar system history. We expect them to be relatively pristine mixtures of ice and rock. The densities of smaller TNOs have been found to be as low as 500 kg/m^3, roughly as dense as wood. These smaller TNOs must thus have significant macroporosity, as is true of smaller asteroids.

COMETARY INTERIORS: "DIRTY SNOWBALLS" OR "ICY DIRTBALLS"?

Comets formed in the same way as asteroids, through accretion of smaller bodies as described in Chapter 5. In the outer solar system, however, accretion speeds were much slower than in the inner solar system, and the planetesimals were much less consolidated than those that formed the asteroids.

Because of this, it is thought that comets are held together much more tenuously than asteroids, and basically consist of many smaller bodies (sometimes called "cometesimals") in a rubble-pile structure. A major difference between cometary and asteroidal interiors is the presence of volatiles like water ice and carbon dioxide ice, where the cometesimals are mixtures of ices and rock. This led Fred Whipple's initial model of cometary interiors to be dubbed the "dirty snowball" model. Later models acknowledged the majority of mass was rock rather than ice, and that an "icy dirtball" was a more accurate description.

Cometary disruptions are relatively common, providing much of our knowledge of their interiors. The most famous of these breakups is probably that of Shoemaker-Levy 9, which was disrupted after it passed too close to Jupiter, and ended up impacting that planet in 1994. It split into roughly 20 pieces. Other comet breakups include Schwassmann-Wachmann 3 in 2006, and a number of "sungrazing" comets. The reason for the disruptions is not certain, and probably differs from object to object. In the case of the sungrazers, it's likely the result of passing close enough to the Sun that tidal effects rip them apart, much like Jupiter did to Shoemaker-Levy 9. For other comets, it's possible that sublimation of volatiles plays a role—as they heat and turn from solid to vapor, the pressure on the interior increases. Because comets are so weak to begin with, this increased pressure could blow them apart.

Shoemaker-Levy 9

The discovery of Comet Shoemaker-Levy 9 (usually shortened to SL9) was a boon to astronomers. The ninth periodic comet found by the team of Carolyn and Eugene Shoemaker and David Levy, its appearance was so unusual that it was not recognized by others who had previously imaged it. Other comets had been seen to be temporarily captured by Jupiter, but none had been seen to have been broken up by a close pass.

Orbital calculations suggested that SL9 was broken up less than a year before its discovery, and more startling, that it would impact Jupiter roughly 16 months after its discovery. Practically every professional telescope on Earth and all available spacecraft were trained on Jupiter during the period of the impacts, including the *Galileo* spacecraft, then en route to Jupiter and possessing a viewing angle unobtainable from Earth—a direct view of the impacts. However, while the impacts themselves were on the far side of Jupiter from the vantage of Earth, the sites rotated into view shortly thereafter and were visible from telescopes as small as 3 cm in diameter. The aftereffects could be seen for months.

While the impacts gave great insight into the atmosphere of Jupiter, it was the pre-impact period that was of most interest for comet researchers. By studying the distribution of SL9's pieces and knowing its orbit around Jupiter, they estimated its original size as roughly 5 km, and found it had only a tiny amount of strength—less than that of snow. It was also determined that Jupiter is likely the victim of a cometary impact every few decades, though one the size of SL9 is much rarer, perhaps once in a thousand years.

Magnetic Fields of Small Bodies

Scientists can gain insight into planetary interiors by studying the magnetic fields they generate. The origin of planetary magnetic fields, and why they have the characteristics they do, is still only poorly understood. However, it is generally agreed that for a planet to have a magnetic field, it needs a layer of conducting fluid. In the Earth, this is the molten outer core, composed of iron.

Given this, we don't expect much in the way of magnetism on the small solar system bodies. However, there are possible exceptions. If Ceres has an internal liquid water ocean, it could conceivably generate a magnetic field. The same is potentially true of other larger objects, like Pluto, although its close cousin Neptune's satellite Triton does not appear to have a strong magnetic field. Vesta is another possibility, and some scientists have recently interpreted color variations on its surface as evidence supporting the presence of a magnetic field on that object. Data from the HED meteorites, originating on Vesta, have been inconclusive.

However, we find that other meteorites are magnetic. Disentangling the history of this magnetism, which diminishes with shock and impact, is difficult, but it is thought that these meteorites were subjected to strong magnetic fields early in solar system history, which "froze in" a magnetic field, just like everyday objects can become magnetized by close exposure to a magnet.

SUMMARY

The interiors of asteroids, comets, TNOs, and dwarf planets vary dramatically in strength. Larger objects have been able to retain radiogenic heat more effectively, and have experienced differentiation as a result, whereupon melting metal, rock, and ice can separate into different layers, with the densest material sinking to form a core and a less dense mantle atop it. Smaller objects were unable to retain heat and stayed undifferentiated. As time went on, impacts fractured smaller bodies, with some impacts large enough to disrupt the target object, which could then reform as a jumbled mass known as a rubble pile. At small enough sizes (perhaps only tens of meters in size), objects cannot reform after a disruptive impact since their gravity is too low, and any objects of that size are thought to be single, coherent rocks. Conversely, at large enough sizes, disruptive impacts are so rare that most bodies have never experienced them, and so objects like Vesta, Ceres, Pluto, and Eris (among others) are also thought to be coherent bodies. While direct evidence about small body interiors only comes from some meteorites at present, we have gained a great deal of knowledge from a starting point of relatively simple estimates of composition in conjunction with measured densities.

WEB SITES

The breakup of Comet Shoemaker-Levy 9 led to a large amount of material being posted on the then-new World Wide Web. A pre-impact summary of what was expected can be found at this Web site. The main page (http://www2.jpl.nasa.gov/sl9/sl9.html) carries updates from a wide variety of sources. Other comet impacts, these into the Sun, are described on http://science.nasa.gov/headlines/y2000/ast10feb_1.htm, along with images of them from the SOHO spacecraft: http://www2.jpl.nasa.gov/sl9/background.html.

This site has HST images and descriptions of the recent cometary breakup of Comet Schwassmann-Wachmann 3: http://hubblesite.org/newscenter/archive/releases/2006/18.

An excellent set of online pages about asteroidal and planetary interiors and differentiation are at this Web site by the American Museum of Natural History: http://www.amnh.org/exhibitions/permanent/meteorites/planets/crust.php.

Astronomy magazine's Web site has a popular-level description of the interior of Ceres: http://www.astronomy.com/asy/default.aspx?c=a&id=3478.

The *Dawn* mission homepage has a set of links about Vesta and its interior: http://dawn.jpl.nasa.gov.

10

Small Body Atmospheres

The small bodies of the solar system are typically thought of as barren rocks hurtling in space. However, while clouds and rain never darken the skies of asteroids, Pluto and other larger TNOs have atmospheres that change and winds that blow. The comae of comets have elements of planetary atmospheres, though with unique aspects. In this chapter, we will look briefly at atmospheres in general before considering the types of atmospheres that are found (or in some cases may possibly be present, if undetected) on the asteroids, comets, and dwarf planets.

SOMETHING IN THE AIR

We are all familiar with the Earth's atmosphere, the part of the planet extending from its solid surface to the edge of space. The atmosphere contains the oxygen that we breathe, hosts the weather systems that bring us rain and snow (in amounts from life-sustaining to life-threatening), and even serves as a form of protection against ultraviolet light and some asteroidal and cometary impacts. Most of the major planets have substantial atmospheres. Some of the small bodies also have atmospheres, although their atmospheres are much thinner than our own and have very different compositions. In this chapter, we will discuss the atmospheres that have been found on comets, asteroids, and dwarf planets, as well as those suspected and those that appear to be absent.

By and large, the more massive a body is, the larger an atmosphere it will have. As an object gets more massive, its gravity increases and it can hold on to gases with smaller molecular weights. In addition, the size of an object's atmosphere is related in a complicated manner to its surface temperature and its distance from the Sun. As temperature increases, a greater number of compounds are found in the gaseous state, enabling a larger atmosphere to be possible. However, at these higher temperatures, on average the molecules in the gas become more energetic and move more quickly. Thus, they are more likely to be moving too quickly to be retained by the object's gravity. The solar distance not only controls the surface temperature but also the compounds available to the body in the first place via the condensation sequence, as was described in more detail in Chapter 5.

The outer bound of a planetary atmosphere is called the **exobase**. Above the exobase, a typical atmospheric molecule is moving at an average speed that is greater than the escape velocity, and so it can escape into space, while below it the molecule is bound to the planet. For most major planets, the exobase is far above the surface—for the Earth, for instance, it is at roughly 500 km, while on Venus and Mars it is over 200 km. For some of the objects discussed in this chapter, the exobase is at the object's surface, meaning any atmosphere we see on those bodies requires constant replenishment since it is constantly escaping.

THE ATMOSPHERE OF PLUTO

When it was first discovered, and it was thought to be Earth-sized or larger, there was a general expectation that Pluto would have an atmosphere. As estimates of its size began to approach modern measured values, however, many began to suspect Pluto to be airless.

The first suggestions that Pluto might indeed have an atmosphere were based on spectroscopic measurements in the 1980s that found evidence for methane in the gaseous phase in addition to solid methane. Thermodynamic arguments were made in light of those measurements, showing that for methane to be present as the spectroscopic results suggested, a second gas, probably nitrogen, was required to make up the bulk of Pluto's atmosphere.

This prediction for an atmosphere on Pluto was confirmed during an occultation in 1988. In Chapter 6, we discussed how occultations are used to determine the size and shape of airless bodies. For objects with an atmosphere, the star being occulted is dimmed by the atmosphere before disappearing behind the solid body. The details of this dimming provide information about the nature of the atmosphere. For Pluto, it was found that the atmosphere was roughly 700,000 times thinner than the Earth's atmosphere. The occultation also provided a measurement of the atomic mass of the molecules in Pluto's atmosphere, confirming the prevalence of nitrogen.

Pluto's eccentric orbit means that its average surface temperature varies considerably. This has led to speculation that it may get cold enough during the plutonian year for the atmosphere to freeze out, leaving the surface coated with frost and ice. During the warmer parts of the plutonian year, the ice would sublime, creating the atmosphere again. Pluto reached perihelion in 1989, and it was thought that atmospheric freeze-out might occur quite rapidly.

However, other occultations of stars by Pluto occurred in 2002, and the results surprised astronomers. Instead of a diminished or absent atmosphere, they showed the atmosphere of Pluto had gotten twice as thick as it was in 1988. This has now been attributed to the changing seasons on Pluto. Shortly after perihelion on Pluto, its northern hemisphere begins autumn and its southern hemisphere begins spring. The south pole of Pluto, in darkness for well more than 100 Earth years, begins to heat up and volatiles that had frozen out on the pole start to enter the atmosphere; as the north pole cools down, it will begin to accumulate volatiles.

However, given our current knowledge about Pluto and accounting for this seasonal effect, scientists now aren't certain whether freeze-out happens or not, though according to some models the plutonian atmosphere may begin to freeze out around the time of *New Horizons*' arrival in 2015. Charon, similar to Pluto in many ways, seems to lack an atmosphere. This is likely because its mass, and thus gravity, is just a bit too small to retain even a tenuous atmosphere.

Even though Pluto has retained its atmosphere over the history of the solar system, molecules are still escaping from above its exopause. It has been estimated that 3–3,000 kg of nitrogen escapes from Pluto every second, or the equivalent of up to several of the Egyptian pyramids' worth of mass per year. If that rate of loss has been consistent since Pluto was formed, this corresponds to between 1 and 10 km of nitrogen ice from its surface that has potentially been lost.

This is an extremely high loss rate, much higher than expected. It potentially tells us much about Pluto's surface. One possible requirement for such a loss rate is that there is not a lot of nonvolatile material mixed with the nitrogen and methane we see on Pluto. If there were sand or rock present, it would gradually come to dominate the surface as the nitrogen escaped. Eventually, no nitrogen would be present at Pluto's surface, and it would only be found beneath a crust, or lag deposit of other material. Some nitrogen might be able to escape by making its way through the lag deposit, but that would only thicken the lag deposit with time. Because we see such a high rate of nitrogen loss at the present day, we can infer that there is no lag deposit present on Pluto.

OTHER POSSIBILITIES FOR TRANSNEPTUNIAN OBJECT ATMOSPHERES

The flood of TNO discoveries in the late-twentieth and early twenty-first centuries postdates the discovery of Pluto's atmosphere. Given what we know about Pluto's atmosphere, it is natural to apply the same logic to the

TNOs and hypothesize which ones may have atmospheres. As with the first predictions for Pluto, this work utilizes the spectral properties of TNOs to figure out their surface compositions. At the temperatures found in the outer reaches of the solar system, water ice is stable even under conditions of very low gravity. Compounds like nitrogen (N_2), methane (CH_4), and carbon monoxide (CO) can be retained by large objects, but not smaller ones.

Eris has a high albedo and evidence of large amounts of methane on its surface. This has been interpreted as evidence for a cycle in which methane frost can condense and sublime, creating a temporary atmosphere as may be the case with Pluto. However, unlike the case for Pluto, it is thought that Eris's atmosphere is currently frozen out on its surface. Sedna, by contrast, has a relatively low albedo. While the data are very sparse and preliminary, this has been used to argue that Sedna does not have an atmosphere at any point in its orbit. Since it is never any closer than 79 AU from the Sun, with a temperature always below 33 K, this is not surprising.

Data for two dwarf planets, Makemake and Haumea, are more ambiguous. Makemake shows evidence of methane ice as well as other compounds that can be created when the solar wind and other energetic photons interact with methane. As with Pluto and Eris, the predicted temperatures for Makemake lead scientists to expect it has an atmosphere during part of its orbit. Interestingly, nitrogen ice is only seen in very small quantities. This has been interpreted as due to preferential escape of nitrogen from Makemake in comparison to other dwarf planets, leaving methane behind.

Haumea shows no evidence of methane or nitrogen at all, and the only compound identified on its surface is water ice. The explanation for this seems to be related to its companions—two satellites and five objects in very similar orbits around the Sun, forming a dynamical family. The origin of the dynamical family is expected to be via a giant impact into Haumea, an impact that may have blown off any methane or nitrogen from the system, leaving only rock and water ice behind. This theory also fits the surprisingly high density of Haumea, much higher than seen for Pluto or Eris.

On objects with tenuous atmospheres like TNOs, small areas of low and high albedos can grow to become large areas. This happens because of the temperature differences that accompany albedo differences—regions with low albedo absorb more sunlight and become warmer. As a result, volatiles in the area experience increased sublimation, and areas adjacent to the higher temperatures also warm up, themselves. Because volatiles tend to have high albedos, increased sublimation leads to lowering the albedo of the warm region. This, in turn, increases the temperature further. The end result can be large low-albedo regions.

However, with the presence of an atmosphere, some volatiles do not escape into space. Instead, the molecules can travel to colder places where they condense as frost (as discussed previously for Pluto). The albedos of the colder areas rise, which, in turn, cools them further in a mirror image of the process for warmer areas. As the seasons change, the warmer and colder regions may move and albedo patterns may change as a result, depending on the amount of the spin axis's tilt (or its **obliquity**). Objects whose obliquity is near zero have little seasonal variation and would end up with permanent high-albedo polar caps.

Putting together the composition of the ices stable on TNOs and their temperatures, scientists estimate that methane atmospheres should be expected within roughly 55 AU of the Sun, while nitrogen atmospheres could exist out to 120 AU. The details depend upon the gravity and density of an object, as well as its obliquity. Regardless of those, however, TNO atmospheres are not expected beyond 120 AU unless something other than solar heating and ice sublimation is creating the atmosphere (such as outgassing due to radioactive heating or impacts adding heat).

COMETS

Although not an atmosphere in the sense usually considered for planets, the coma of a comet fits the definition of an atmosphere. As with some of the TNOs that we have discussed, comets have volatile ices that are stable as solids when they are further from the Sun, but become gases as they warm up closer to the Sun. The coma is composed of these gases. In an important difference with KBOs, however, the materials in cometary comas (or alternately, comae) are not recycled from season to season but rather escape, with new gas and dust released during each orbit.

Cometary comas are most well-developed for comets nearer the Sun. While the TNO atmospheres mentioned previously are composed of gases like methane or nitrogen, comets near 1 AU have temperatures warm enough to sublime water ice, which is by far the most abundant of the ices. The coma size is much larger than the nuclear size, in contrast to the atmospheres of the terrestrial planets, which are very thin compared to the planetary radius. In addition to gas, the coma contains dust lofted from the nucleus by the escaping gas.

As the gas in the coma expands, it begins to cool rapidly, reaching a minimum temperature of only 30–40 K. As it expands, however, the lofted dust is less able to shield it from sunlight, which eventually begins to reheat it. In addition to heating, energetic ultraviolet photons from the Sun **dissociates**, or breaks apart, the original gas molecules in the coma into smaller forms. These **daughter molecules** can be detected spectroscopically and give information about their original parent molecules as well as the rate of gas outflow from the nucleus. For instance, a molecule of water (H_2O) can dissociate into OH and H, or H_2 and O. The daughter molecules can often themselves be broken down as well, in this case the OH could further dissociate into O and H. In addition to dissociation, ionization can occur, leaving OH^- or H_2^- ions in the coma.

Comets making their first visits to the inner solar system ("new" comets) are often extremely active. In the transneptunian region, their temperatures are 30 K or lower. In the Oort cloud, they are lower still, perhaps colder than 10 K. As they approach the Sun and heat up for the first time in billions of years, there is plenty of material to be blown off. They experience, in effect, a reversed version of the condensation sequence described in Chapter 5—the most volatile ices begin subliming farthest from the Sun, followed by water ice sublimation. "Older" comets, those that have spent a longer time in the inner solar system, have largely lost their volatile ices and have comae dominated by water, relatively speaking.

Because it takes time for heat to penetrate into the interior of an object, and because new comets have a large reservoir of volatiles near their surfaces, activity can sometimes be seen for new comets even when they are past perihelion and heading farther from the Sun. Comet Hale-Bopp, for instance, still had a visible coma and tail even a decade after its perihelion and after it had retreated past the orbit of Uranus (20 AU from the Sun).

Breakups of comets can also be responsible for new activity. Comets Shoemaker-Levy 9 (SL9) and Schwassmann-Wachmann 3 (SW3) both turned into chains of comets as they broke into their constituent pieces and parts of their interiors saw sunlight for the first time. The composition of

Figure 10.1 When Comet Schwassmann-Wachmann 3 (SW3) broke up, fragments of its interior were exposed to sunlight for the first time. This caused ice to quickly sublime and create a coma and tail for each fragment. As a result, each fragment became a tiny comet of its own. In this infrared image from NASA's Spitzer Space Telescope, dozens of fragments are seen, each with its own tail. AP Photo/NASA.

SW3's coma after it broke up was compared to its pre-breakup coma, and was found to be the same, suggesting its interior composition was similar to its former surface composition. This is evidence that the surface of SW3, and by extension, most comets, doesn't change much with time.

TALES OF TAILS

Cometary tails can properly be considered as the result of coma evolution. The gas and dust present in a coma both interact differently with the forces present in interplanetary space. The dust grains are small enough that they can be affected by a force called radiation pressure. Radiation pressure is caused by a transfer of energy from photons striking a surface. It is an extremely weak force and is easily neglected in everyday life. It is strong enough, however, to affect the tiny grains of dust and overcome the weak gravity of a comet. Once lofted into the coma, radiation pressure acts on the dust grains to push them away from the Sun. Without the effect of radiation pressure, cometary dust would remain in an orbit around the Sun very much like their parent nucleus. When adding radiation pressure, their paths change so as to stretch out behind the nucleus, along the cometary orbit. This is why **dust tails** have the appearance they do.

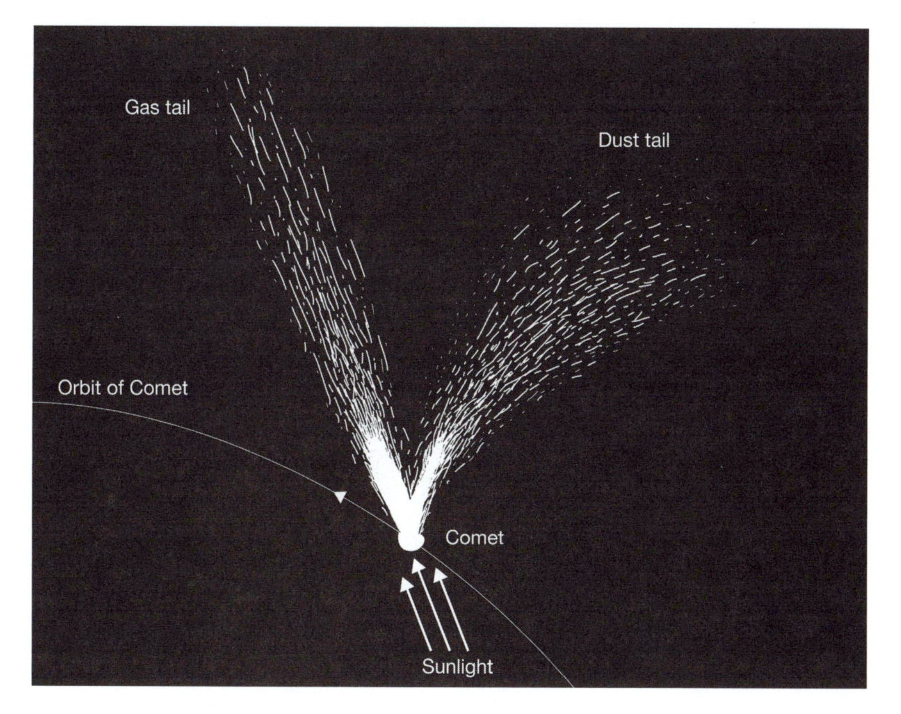

Figure 10.2 The gas and dust in a cometary coma act differently as they are swept out into the tail. The dust follows separate orbits around the Sun under the influence of its gravity while being pushed by sunlight, while the gas is influenced only by the pressure of sunlight and moves directly away from the Sun. This results in two tails for most comets, as seen in this image. Illustration by Jeff Dixon.

Comets also have another kind of tail. The gas in a cometary coma is not affected by radiation pressure. Ionized gas can be affected by other forces, however. The solar magnetic field creates a force felt by charged particles, which can be stronger than gravity. The ionized gas ejected by a comet moves in response to the solar magnetic field, creating an **ion tail** or **gas tail**, which typically is directed away from the Sun. The ion and dust tails of a comet may be aligned with one another to a greater or lesser extent, but they are commonly separate, and differences in color can be seen by a sharp-eyed person.

CHIRON AND CENTAURS

The Centaur Chiron is in many ways the largest known comet. It was first discovered to have a coma in 1989, and its brightness has varied with time, suggesting outbursts of activity similar to comets. As a result, it was given a cometary designation (95/P Chiron) to go along with an already-given asteroid number (2060). Therefore, it is one of the few objects classified as *both* a comet and an asteroid.

After its coma was detected, Chiron was observed frequently through its perihelion in 1996 and beyond. Confusingly, the observed outbursts have not been correlated with distance from the Sun (and presumably temperature) in a consistent way. Water ice has been detected on Chiron's surface, however, even at its hottest surface temperature it will remain solid. As was discussed in Chapters 8 and 9, the sublimation of near-surface water ice is thought to be responsible for most cometary activity. CO and CN have been detected in Chiron's coma, and have much lower sublimation temperatures—as low as 25 K for carbon monoxide. Theoretical studies suggest that the sublimation of CO may be responsible for Chiron's activity. An additional possibility is that temperature-dependent changes in the type of water ice in Chiron can lead to outbursts. As Chiron moves toward aphelion in 2021, it will continue to be monitored. In addition to Chiron, there is one other Centaur known to have a coma, 2001 T4. Little else is known about that object at this time.

MAIN-BELT COMETS

As discussed in Chapter 9, it is thought that many asteroids could have significant amounts of water ice in their interiors. In theory, that ice could sublime if it became warm enough, just as newly exposed ice on a cometary surface warms and gives rise to a coma.

In 2006, researchers at the University of Hawaii discovered a handful of objects on asteroidal orbits that have cometary tails and appear to have comae. Dubbed **main-belt comets** or **activated asteroids** because of their nature, these objects are the best evidence that comets and asteroids grade into one another rather than being compositionally unrelated. The known main-belt comets are members of the Themis dynamical family, found roughly 3.1 AU from the Sun. While they are too small for detailed compositional study to be done at present, as members of the Themis family, they would be expected to be low albedo, C-class asteroids analogous to carbonaceous chondrite meteorites (see Chapters 4 and 7). The comae and tails of these objects appear to be transitory, consistent with an impact excavating an ice layer from beneath the surface, which then moves in and out of sunlight with changing seasons, corresponding to times when a coma is seen or absent. As discussed in Chapter 8, a lag deposit would be expected to form on main-belt comet surfaces just as on comets. Within a few years, we will know what fraction of Themis family objects are activated asteroids. Using that fraction, and estimates of how quickly lag deposits form, scientists will be better able to understand how frequently impacts occur in the Themis family.

CERES

Interestingly, Ceres has also shown evidence of outgassing. Spectroscopic observations of Ceres in the early 1980s suggested that there might be water ice on its surface, though at the temperature expected on Ceres, it would not be stable for long periods. This led to a search for evidence of ice sublimation using the *International Ultraviolet Explorer* (IUE) satellite. The results were not definitive, but hinted that OH might be present near the pole of Ceres then experiencing summer, while the other pole showed no such excess of OH. This was interpreted as consistent with ice heating up and subliming off of the summer pole, then dissociating into OH and hydrogen. However, this observation has never been confirmed.

Subsequent findings about Ceres have not favored the presence of water ice at its surface, as discussed in Chapter 7. However, the discovery that Ceres may have a vast internal ocean of ice opens the possibility that water could potentially be outgassing and Ceres might have a tenuous atmosphere. The definitive word on an atmosphere for Ceres will likely need to await the arrival of the *Dawn* spacecraft in 2015.

TYPICAL ASTEROIDS

The main-belt comets and Ceres are unusual asteroids because they have near-surface ice that can sublime to create an atmosphere, like the TNOs discussed earlier. The vast majority of asteroids have rocky surfaces, which are stable and do not sublime. Other than the objects mentioned previously, therefore, no main-belt asteroids have been seen or suspected to possess an atmosphere. However, it is possible that some of the largest ones like Vesta could in theory have what is called a ballistic atmosphere, where the density of molecules is so low that they are much more likely to hit the surface or escape than they are to encounter another atmospheric molecule. This type of atmosphere is found on Mercury, the Moon, and several large planetary satellites. In those cases, the atmosphere is thought to originate from the solar wind knocking atoms like sodium or potassium out of surface rocks (or knocking a hydrogen or oxygen atom out of an ice molecule on an icy body). Those atoms can escape quickly or bounce around a few times, depending upon their energy. Another possible origin for these atmospheres is outgassing of material from an object's deep interior, as could be occurring with ice on Ceres. The total atmospheric pressures for these atmospheres are very small, about 10^{12} times less dense than the surface pressure here on Earth. Such an atmosphere would be difficult to observe using ground-based equipment, but observations from Vesta's orbit might be able to detect one. Smaller objects do not have gravity strong enough to maintain even these exceedingly thin atmospheres.

SUMMARY

Unlike the planets, the small bodies of the solar system do not have thick atmospheres. However, there are thought to be thin atmospheres on several of the large TNOs. The only one of these TNO atmospheres directly detected is Pluto's, which is generated and maintained by sublimation of ice from its surface. Our expectations of other TNO atmospheres are based on comparison to Pluto and theoretical calculations. For some objects, the presence of an atmosphere is indirectly inferred. The sublimation of ice from comet surfaces creates a temporary atmosphere, which we see as comae and tails. The gas and dust in these atmospheres escape, requiring further replenishment. Asteroids do not have atmospheres in general, although a class of "main-belt comets" has comae and tails like normal comets, indicating the presence of near-surface ice. The study of small body atmospheres gives insight into the conditions at their surfaces.

WEB SITES

A more detailed discussion of Pluto's atmosphere, touching on other TNO atmospheres, can be found at this Web site, which also has sections focusing on small body satellites: http://www.johnstonsarchive.net/astro/pluto.html.

Dr. Anne Sprague maintains a set of information about the atmosphere of Mercury at this Web site. Although it is not an asteroid, Mercury's atmosphere is similar in character to what one might hypothetically expect on Vesta, if conditions were right: http://www.lpl.arizona.edu/~sprague/planatmos/mercatmos.html.

The codiscoverer of the first Kuiper belt object, Dr. Dave Jewitt, has written a set of Web sites about comets and TNOs, including two focusing on the coma and tails: http://www.ifa.hawaii.edu/~jewitt/coma.html and http://www.ifa.hawaii.edu/~jewitt/tail.html.

11

Small Bodies and Hazards

While most of the small bodies of the solar system are hundreds of millions of kilometers or more from the Earth, some have orbits that bring them much, much closer. Occasionally, the Earth is hit with something large enough to penetrate the atmosphere, sometimes with dire consequences for life on this planet. In this chapter, we discuss the hazard posed to civilization by asteroid and comet impacts, and the initial steps humanity has taken to try and understand and defend against this threat.

THE IMPORTANCE OF IMPACTS

The Earth has been peppered by impacts throughout its history. While erosion by weather and tectonic processes have erased most of the craters it has accumulated over billions of years, over 170 craters with diameters from 300 km down to a few dozen meters can still be found. Roughly 54 tons of mass falls on the Earth per year, with objects up to 3 m diameter falling in a typical year. The largest impactor in historic times occurred in 1908 over Siberia, an airburst of a 50-m-diameter object. For comparison, the Sikhote-Alin fall of 1947, the largest meteorite fall recorded, involved an iron impactor of only 10 m diameter—roughly half the size of a railroad car. Typically, chondritic impactors smaller than 100 m or so do not reach the ground intact, and objects only a few meters across or smaller completely burn up in the atmosphere. As objects get larger (or denser), the atmosphere has less ability to slow them down. Impactors of 100 m or larger hit the ground in **hypervelocity impacts**, with speeds faster than the speed of sound in rock. Impacts at these speeds have more in common with explosions than a rock dropped from a skyscraper, or even from an airplane. The impactor often vaporizes, leaving little of it behind.

The consequences of impacts were first realized in the 1980s. One contributing factor was research into "nuclear winter," which was a hypothesized change in climate that could be caused by a large-scale nuclear exchange between the United States and Soviet Union. The possible climate change would be driven by large amounts of smoke and soot blocking sunlight from reaching the surface. It was realized that these same conditions could be created by an asteroid or comet impact rather than a nuclear exchange.

The most influential research in focusing scientific attention on NEO impacts, however, was the proposal that such an impact was responsible for the extinction of the dinosaurs. This occurred as the Cretaceous Period ended and the Tertiary Period began, called the **K-T boundary** in the rock record (the traditional abbreviation for the Cretaceous is K). The proposal that the extinction was caused by an impact was based on an iridium-rich layer of rock found at the boundary. As discussed in Chapter 4, iridium is a siderophile element and is concentrated in the Earth's core, rarely found in surface rocks. As a result, the abundance of iridium in this rock layer was interpreted as extraterrestrial in origin, brought in by an impactor. Soon after, a crater of the correct age was found on the Yucatan Peninsula of Mexico.

The recognition that impacts can lead to extinctions helped spur the first dedicated searches for near-Earth objects. These searches, discussed in more detail later, along with modeling of impacts, have provided a general sense of how often collisions occur with various-sized NEOs, and how bad those impacts are for life on our planet.

While large enough impacts can devastate life on Earth by changing the climate and starting wide-ranging fires, smaller impacts can also potentially have effects that cause regional destruction. Small impacts occur much more frequently than large impacts, although very small objects burn up in the atmosphere and are not dangerous. Objects a few meters in size reach the ground as meteorites, but are slowed by the atmosphere to the point that they are no longer hypervelocity impactors, and do no more damage than a rock dropped from a great height—potentially bad for any unlucky person who might be struck by one (or who owns a car or house struck by one), but not a general problem for the community.

An important facet of current research is trying to determine more precisely the size at which impactors become hazardous. This is complicated by the fact that scientists are still working on the effects of impacts into water. The most worrisome aspect of ocean impacts is the possibility for a tsunami, which could affect large areas, some of which are heavily populated. Computer simulations of tsunamis are generally not designed to study impact-generated events, which can have very different characteristics than those generated by earthquakes. In addition, most existing simulations consider those areas where tsunamis have commonly been seen (for instance, the Pacific or Indian oceans, starting along earthquake faults), while an impact-generated tsunami could theoretically happen anywhere, including the Atlantic Ocean or even the Mediterranean Sea.

In September 2007, a crater-forming impact on Earth was recorded for the first time in history. Near the town of Carancas, Peru, a fireball was followed by an explosion that broke windows 1 km from the impact site, leaving a crater 5 m deep and 13 m across. While the original size of the impactor was relatively small, the high elevation at the impact site (over 3,800 m) meant that the screening effects of the atmosphere were not as large as they are at sea level.

Remarkably, only a year later, the first Earth impactor to be discovered prior to impact was found—2008 TC3, an object about the size of a compact car. Roughly a day before impact, 2008 TC3 was discovered, and it was quickly realized that it would impact near the Egyptian-Sudanese border carrying roughly a kiloton of TNT's worth of energy, creating an extremely bright **bolide** (or fireball) upon entry. Scientists were uncertain if pieces of 2008 TC3 would reach the ground as meteorites, and because of the remote impact location, recovery of any such meteorites would be difficult. However, meteorites were successfully recovered in early 2009, and analyses are under way.

The Tunguska Event

The largest impact in recorded history occurred 7:40 AM on June 30, 1908, in Siberia. For decades, however, nobody quite knew what had happened. Seismic stations throughout Europe recorded the blast, and the fires that resulted put enough smoke in the atmosphere to be noticed by North American observatories for several months.

The remote location of the impact in combination with the upheaval that plagued Russia for the following decades meant that the blast site could not be visited for some time. When the first scientific expedition reached the Tunguska site in 1927, they found an area of devastation 50 km in diameter, though interestingly no crater. Studies of the impact site show the blast had a power of several megatons, as much as 1,000 times the power of the Hiroshima atomic bomb. It is fortunate that the impact occurred in remote Siberia, rather than over a city, which would have been a catastrophe.

Before the physics of impacts were well-understood, it was not recognized that an asteroid impact could have such devastating results, and well into the late twentieth century alternate explanations were concocted for the "Tunguska Event," as it was called, including collision with a "mini black hole" or a crash of an extraterrestrial spacecraft. Consensus now has settled on the more prosaic explanation of an asteroid (or, less likely, comet) impact. Impacts the size of the Tunguska Event are expected on Earth roughly every thousand years.

Figure 11.1 Tunguska site. AP Photo.

MEETING THE NEIGHBORS

The first near-Earth object (NEO) to be discovered was 433 Eros, in 1901. At the time, its orbit was seen as a curiosity, though it was recognized that Eros's proximity to the Earth could be used to better calculate the distance between the Earth and the Sun. Additional NEOs were found over the next several decades, though the first dedicated searches for them were not begun until the last quarter of the twentieth century.

The passage of 4581 Asclepius by Earth in early 1989 at roughly twice the distance of the Moon spurred studies of how to determine the near-Earth object population. The potential consequences of an impact were dramatically shown in 1994 when Comet Shoemaker-Levy 9 hit Jupiter. This set of impacts resulted in spots visible in Jupiter's atmosphere for months afterward in even small telescopes, underscoring the energy involved in the impact of even small objects.

In 1998, the United States Congress mandated NASA to find 90 percent of all NEOs with sizes 1 km diameter or larger, capable of causing global catastrophe, within 10 years. NASA has funded the Spaceguard Survey as a result, composed of several independent surveys, each with specific strengths. As of the closing months of the Spaceguard Survey, astronomers estimate they will come close to surveying the NEO population to the requested level. The first searches used technology similar to that used to discover Pluto in the 1930s—photographs of the same patch of sky taken at different times and visually compared to one another using a blink comparator, which is a machine that rapidly switches back and forth between two photographs as seen through an eyepiece. Objects like stars remain fixed between the two frames, while moving objects jump back and forth between their positions on each photo. This technique is still used with digital imagery on computer screens, where it is called **blinking**. Newer search programs developed software that could automatically identify every object on an image and compare it to images of the same area at different times. After removing all of the objects whose positions are identical on the different images (and thus are fixed stars), the remaining objects are possible NEOs (or main-belt asteroids or comets). Current techniques require at least three images to be observed to allow the software to determine if any potential NEOs are moving in a manner consistent with a real orbit.

While different search programs have differing strengths, all must deal with certain observational limits and difficulties. First is that NEOs can go through phases, just like the Moon. As a result, they are brightest when their phase is "full" and the Earth is directly between them and the Sun. This alignment is called **opposition**. Further from opposition, objects become fainter. Objects at opposition are always highest in the sky at midnight, opposite the Sun. Many search programs therefore concentrate on that part of the sky, knowing that any NEOs they find will be nearly as bright as they can be.

Light from the Moon interferes with observing nearby. As with NEOs, full moon occurs at the opposition point in the sky, and observing fainter objects nearby becomes difficult to impossible. As a result, many search programs choose not to observe at all near the time of full moon. The Milky Way is also a difficult area to observe, due to the very high concentration of stars and the likelihood of confusing a faint object with a background star.

Distant objects like Eris and Pluto move so slowly that their opposition dates change very slowly from one year to the next. Objects closer to the Earth's orbit have opposition dates that change more rapidly. The time between oppositions is called the **synodic period**. This can be thought of as though the Earth and other solar system bodies are cars on a racetrack. The synodic period is how long it takes Earth to lap other objects, and the more similar an object's speed is to the Earth's, the longer it takes Earth to lap them. The synodic period of many NEOs can be very long, perhaps decades. Because the objects with orbits most similar to the Earth are also the ones most likely to impact the Earth, and because their synodic periods can be so long and they spend so much of their time far from opposition, some search programs have begun to include search areas far from the opposition point.

Figure 11.2 shows a simplified version of the situation for Earth and the NEO 10563 Izhdubar, with the relative positions of Earth, the Sun, and Izhdubar projected onto the ecliptic plane on March 21 of every year. Objects to the right of the vertical line are visible during the night, and objects to the left are visible during the day, though close-to-the-line objects can be visible at night close to dawn or dusk. In 1993, Izhdubar was discovered, and in 2009–2010 it will be opposite the Sun from Earth. Izhdubar's year is roughly four days longer than the Earth's and it will slowly fall behind the

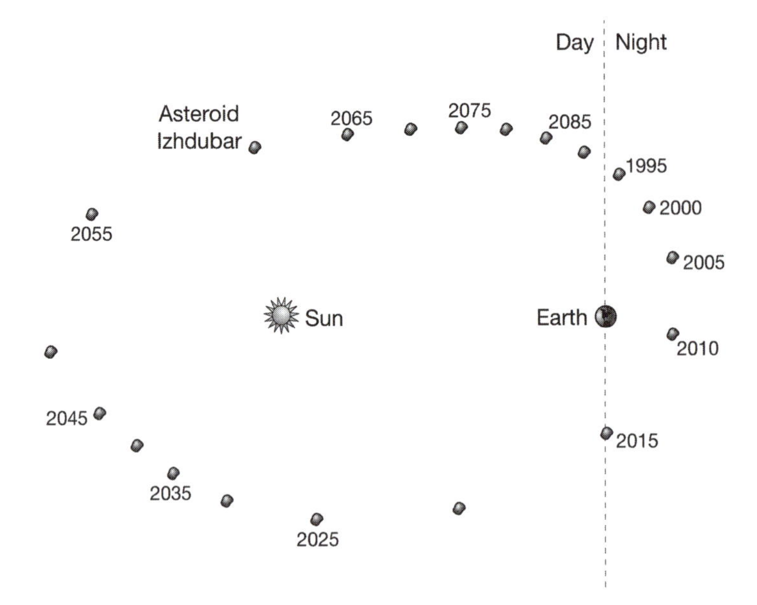

Figure 11.2 This diagram is a snapshot of the positions of the Earth, Sun, and the asteroid 10563 Izhdubar every fifth March 21 from 1995 to 2090. Because it is the same day each year, the Earth is always in the same spot relative to the Sun. Izhdubar, however, is found in very different parts of the sky on March 21 as the years pass. The vertical black line separates the times when Izhdubar is visible during the daytime (to the left of the line) and at night (to the right of the line). As can be seen, for the vast majority of the 100-year period shown, Izhdubar is not visible at night. The same is true of many other NEOs, which has led search programs to try and counter this effect by proposing space-borne telescopes which could search much closer to the Sun. Illustration by Jeff Dixon.

Litter in Near-Earth Space

The vast majority of material outside Earth's orbit is transported from the main asteroid belt. However, humanity has sent a number of space probes to explore the solar system. While the main probes are tracked, booster rockets, instrument covers, and the like are often jettisoned and untracked (or untrackable). As NEO surveys become sensitive to smaller and fainter objects, some of these objects will be detected and catalogued, though definitive identification could be difficult.

A dramatic example of this was seen in 2007 as the European *Rosetta* spacecraft was preparing for a gravity assist using the Earth. Ground-based surveys identified it and it received a provisional designation of 2007 VN84 before suspicious observers noted the similarity between its orbit and *Rosetta*'s. In 2002, an object with an orbit very similar to Earth's was discovered and named J002E3. Astronomers (including your humble author) obtained its reflectance spectrum, and found that it matched white titanium-based paint, clinching a human-made origin. This was supported by orbital calculations that showed J002E3 had last made a close approach to Earth in 1971. The current best guess for the identification of J002E3 is an upper stage for Apollo 12. J002E3 is once again wandering in interplanetary space, and is expected back in the neighborhood in roughly 2032.

Earth as they both make their way around the Sun. For most of Izhdubar's orbit, it is only visible during the day, spending decades at a time close to the Sun (for instance, between 2030 and 2060). However, it also spends several decades visible shortly before sunrise (for instance, the period from 2075 to 2095 and beyond). This type of behavior is true in general for NEOs, not just Izhdubar. These areas near sunrise and sunset have informally been dubbed "the sweet spots" by asteroid searchers and are prime targets areas for the newest NEO searches. The sweet spots are particularly important when observing those objects whose orbits are largely or entirely within the orbit of the Earth. Such objects spend very little time visible in the nighttime sky, making them particularly difficult to discover.

UNDERSTANDING AND COMMUNICATING THE NEO HAZARD

Our current understanding of the NEO population and its threat is shown in Figure 11.3. There are estimated to be roughly one thousand objects of the size capable of causing global devastation with an impact. None of the ones we know of are on orbits that will impact the Earth. The size-frequency distribution (a concept first discussed in Chapter 6) of NEOs leads us to expect roughly 50,000 objects larger than 140 m, which is a size that could cause regional devastation, with an impact energy the equivalent of a

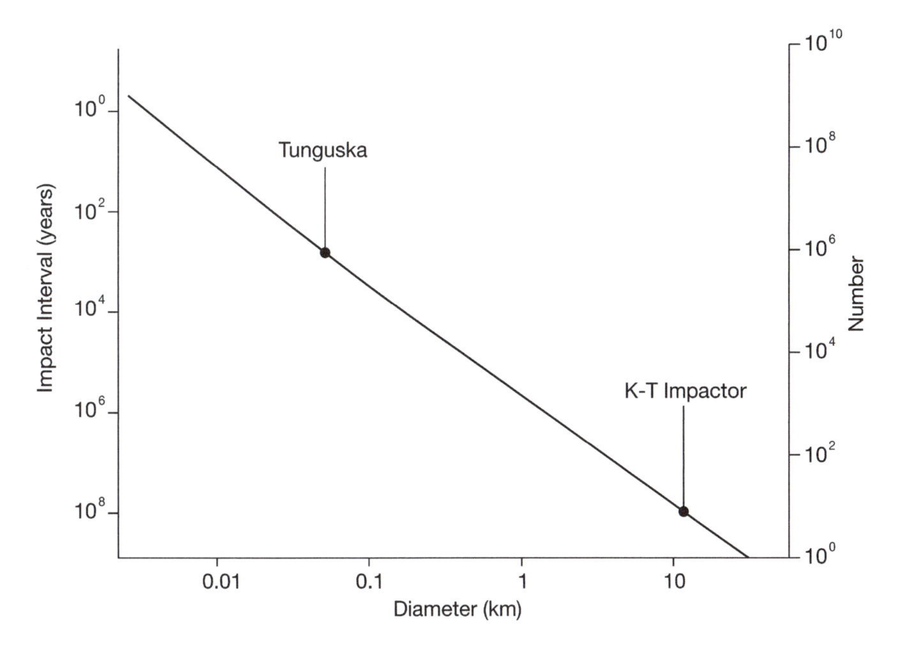

Figure 11.3 The size frequency distribution of NEOs, in combination with the average speeds with which they impact the Earth, can be used to estimate the frequency of impacts of different sizes. Larger impacts occur rarely but have devastating consequences. Illustration by Jeff Dixon.

typical nuclear weapon (5–10 megatons of TNT). As mentioned before, smaller objects do not usually make it through the Earth's atmosphere unless they are the relatively rare objects made of iron. The calculations used in the figure are based on objects with the density of rock rather than ice or metal. The denser the impactor, the bigger the energy of the impact and the more destructive it is. An impact speed typical of asteroids is also used.

The role of comets in the impactor population is not completely understood. They are believed to be a relatively infrequent threat compared to the asteroids. In addition, searches for hazardous comets are beyond the capabilities of current search programs. A cometary impact is likely to be at extremely high speed, potentially up to four times faster than a typical asteroidal impact. As a result, the warning time might be exceedingly short. While cometary impacts are currently not considered to be preventable, it is likely that within the next few decades the largest remaining uncharacterized risk will be from their potential.

Given the large number of potentially hazardous asteroids and the fact that orbits are not immediately known to high precision, it is not surprising that some objects have a non-zero impact probability shortly after their discovery. Because this is an unfamiliar threat to many, coverage of these situations in the media has occasionally been confused.

The increasing number of NEOs that have been discovered have led to a corresponding increase in the number of possible impactors that have been found. Additional data for these objects, whether newly acquired or archival, has always improved knowledge of their orbits to allow an impact to be ruled out. However, before this additional data has become available, the threat of impact is not zero. The most dramatic case of this was in late 2004, when the object 99942 Apophis, then known by its provisional name of 2004 MN4, was thought to have a 1 in 38 chance of hitting the Earth in 2029 before new data showed there would be no impact. There have been some cases where the media has publicized a possible impact while the calculations were very uncertain and additional data were being gathered. Scientists have weighed the public right to know about any potential collision with the desire to present a complete, considered picture to the public and avoid any panic. And unsurprisingly, differing personalities lean toward different ends of the argument.

Part of the effort to better inform the public of the risk of NEO impacts has been the creation of two measures of the threat from any specific asteroid. The **Torino Scale** (Figure 11.4) categorizes objects based on the likelihood of impact and the consequences if an impact occurred, ranging from 0 for objects that will not hit the Earth or will burn up in the atmosphere, all the way to 10 for objects that are civilization-threatening. Apophis reached 4 on this scale before new data allowed it to be recategorized as a 0. The object 2008 TC3 was sufficiently small that even though it impacted the Earth, it never reached more than 0 on the Torino Scale. The **Palermo Scale** has been proposed as an alternative to the Torino Scale. It is more

Category	Scale	Description
No Hazard (White Zone)	**0**	The likelihood of a collision is zero, or is so low as to be effectively zero. Also applies to small objects such as meteors and bodies that burn up in the atmosphere as well as infrequent meteorite falls that rarely cause damage.
Normal (Green Zone)	**1**	A routine discovery in which a pass near the Earth is predicted that poses no unusual level of danger.
Meriting Attention by Astronomers (Yellow Zone)	**2**	A discovery, which may become routine with expanded searches, of an object making a somewhat close but not highly unusual pass near the Earth.
	3	A close encounter, meriting attention by astronomers. Current calculations give a 1% or greater chance of collision capable of localized destruction.
	4	A close encounter, meriting attention by astronomers. Current calculations give a 1% or greater chance of collision capable of regional devastation.
Threatening (Orange Zone)	**5**	A close encounter posing a serious, but still uncertain threat of regional devastation. Critical attention by astronomers is needed to determine conclusively whether or not a collision will occur.
	6	A close encounter by a large object posing a serious but still uncertain threat of a global catastrophe. Critical attention by astronomers is needed to determine conclusively whether or not a collision will occur
	7	A very close encounter by a large object, which if occurring this century, poses an unprecedented but still uncertain threat of a global catastrophe.
Certain Collisions (Red Zone)	**8**	A collision is certain, capable of causing localized destruction for an impact over land or possibly a tsunami if close offshore. Such events occur on average between once per 50 years and once per several 1000 years.
	9	A collision is certain, capable of causing unprecedented regional devastation for a land impact or the threat of a major tsunami for an ocean impact. Such events occur on average between once per 10,000 years and once per 100,000 years.
	10	A collision is certain, capable of causing global climatic catastrophe that may threaten the future of civilization as we know it, whether impacting land or ocean. Such events occur on average once per 100,000 years, or less often.

Figure 11.4. The Torino Scale is an attempt to inform the public about the hazards posed by specific objects based on their odds of impact and the energy that would be released if they did impact (which is related to their size). As orbital information improves for an object, it can move between categories. The bottom panel provides an explanation of each category. Every known object is either a 0 or 1 on the Torino Scale. The NEO 99942 Apophis briefly reached a 4 on the Torino Scale before additional data moved it back to 0. This was the highest any object has yet reached. NASA.

technical, and includes the time until possible impact as part of the calculation. While the Torino Scale only includes integers, the Palermo Scale allows for fractions. The Palermo Scale also includes a comparison to the "background" probability of an impact, which is the chance Earth will be hit by something currently undiscovered. Again, Apophis has had the largest Palermo Scale value in history, topping out at 1.1 before falling to −2.5. Because the Palermo Scale is a logarithmic scale, this is read that Apophis at its peak was $10^{1.1}$ (or about 13) times more likely to impact than a currently undiscovered object, but now is only $10^{-2.5}$ or roughly 1/300 as likely.

Hazards in Space

Impacts do not only occur on Earth, of course. Much of our knowledge about impacts comes from studying their effects on the Moon and other planets. The only large impact ever witnessed was that of Comet Shoemaker-Levy 9 into Jupiter during the mid-1990s, though for some time in early 2008 while its orbit was uncertain it was thought that the NEO 2007 WD5 had a chance of impacting Mars.

Until relatively recently, humanity has not had to worry much about non-Earth hazards. With the International Space Station (ISS) constantly staffed with astronauts and cosmonauts and the possibility of lunar bases in coming decades, however, scientists have begun to calculate the likelihood of impacts affecting them.

Unlike people living on Earth, space stations and lunar bases do not have the benefit of an atmosphere to screen out small, fast-moving micrometeorites. Furthermore, scientists do not have a precise idea of how many of the smallest micrometeorites exist in space. And the consequences of even tiny impacts could be severe if they are unluckily placed or puncture pressurized areas. There have been some events, for example, the old Soviet/Russian Mir space station suffered four small impacts in 15 years, though none were serious. A study conducted in 2007 estimated that the ISS had as much as a 5 to 9 percent chance of being destroyed by a meteoroid impact.

Lunar bases under consideration also run similar risks. However, they have some potential advantages that space stations don't have. The most obvious advantage is that they can be built partially or wholly underground, which would protect them from all but relatively uncommon large impacts. Conversely, the cheapest and fastest construction for a lunar base would be using inflatable habitats, which would be much more susceptible to micrometeorite damage than the ISS.

NUDGE OR SMACK?

Along with the efforts to catalogue the near-Earth and potentially hazardous object populations, there has been research into the best way to handle an object on a collision course. These **mitigation** techniques vary from relatively gentle to more violent. Not surprisingly, the gentler techniques require longer times to be effective. At the most basic level, mitigation ideas can be separated into impulsive techniques, which attempt to deflect the incoming object with a single blow, and continuous techniques, which nudge the target without running the risk of disrupting the target. The most efficient way of changing the would-be impactor's orbit is by slowing or speeding it up—the Earth moves a distance equal to its radius in under four minutes, so a change in arrival time by that amount should allow an impact to be avoided.

The most dramatic of the impulsive techniques is the use of a nuclear weapon (or several nuclear weapons) to deflect the target. Contrary to the conception of this method often shown in movies, the weapons would not be used to blow up the incoming body, which could potentially lead to multiple collisions by the resulting fragments and leave us worse off than originally. Instead, the nuclear weapons would be detonated at some distance from the body, and the absorption of neutrons from the explosion would impart momentum to the NEO and move it.

Many people are exceedingly uncomfortable with the use of nuclear weapons for mitigation, even for the cause of averting catastrophe. There is some fear that acceptance of their use, even in a good cause, would lead to the militarization of outer space. Others argue that a large number of launches may be required to ensure that an incoming object is deflected, and the consequences of a launch failure for a nuclear weapon could be devastating. Nevertheless, the general consensus among researchers is that if a large object is found a short time before impact in the near future, deflection via nuclear weapons may be our only realistic response.

The other major impulsive technique is use of a so-called kinetic impactor. A kinetic impactor is simply a mass sent to collide with a target with a speed and direction to impart the required momentum to move the target away from a collision course. It is effectively the same technique used in billiards. Kinetic impactors are attractive because they could provide a straightforward means of deflection that does not use nuclear weapons. However, the details could be difficult to work out since unlike billiard balls, NEOs will create ejecta when they are hit, which also will carry momentum and affect the final trajectory of the target. Further research has been proposed to better understand how kinetic impactors can be used, and "practice missions" have been suggested as a means to do so. It is likely that this technique will be the first one tried out as a test.

Continuous techniques cause small changes in target velocity at any given time. They are considered to be better suited to smaller objects, and particularly objects that do not require large orbital changes to miss the Earth, particularly objects for which potential impacts are far in the future. Continuous techniques have the benefit of allowing constant monitoring of the impactors' orbits, without any of the uncertainties associated with impulsive techniques. Conversely, continuous techniques can potentially require spacecraft to operate for very long times, and a failure partway through a mission may end up simply moving an impact to a different location on Earth. Finally, continuous techniques require a rendezvous with the NEO to be deflected, while impulse techniques can potentially be used on less-expensive flyby missions (although targeting on flyby missions will potentially be challenging).

Perhaps the most obvious of the continuous techniques is literally pushing the impactor out of Earth's way. In the simplest case, this would involve a spacecraft attaching to a NEO and firing a rocket. This type of mission, sometimes called an "asteroid tug," would need to account for the rotation of the target, to not simply increase its spin rate rather than moving it. Proponents of the asteroid tug have suggested that a few months of continuous operation roughly a decade before any potential impact date would be sufficient to deflect a typical threat.

The mass of an impactor can be estimated given a density and size. For a 100-m non-porous rocky body, this works out to roughly 1.5 billion kg. For comparison, the Space Shuttle orbiters have a mass of 100,000 kg. While this is only a small fraction of the NEO's mass, if the Space Shuttle were close enough to a 100-m NEO, it would exert a gravitational acceleration on it. With enough time, a threatening NEO could in theory be deflected simply by putting a massive spacecraft nearby and "dragging" the NEO using the spacecraft's gravity. This concept, called the "gravity tractor," has been proposed by some as an ideal way of handling threats from small NEOs. It has the benefit of not requiring any attachment to the target and being insensitive to the rotation rate or spin direction, in contrast to the asteroid tug. However, it requires the tractor to be in operation at a very small distance from the target, which might be difficult to manage. The gravity tractor is another concept that may be tested in the relatively near future.

Additional mitigation ideas have been proposed, but some initial studies show them to have greater issues than the ones already discussed. The most notable of these ideas are the use of a "mass driver" and changing the object's albedo. A mass driver removes material from the target surface and ejects it at high speed. This, due to conservation of momentum, would slowly propel the object in the desired direction. However, rotation of the target must be dealt with in the same way as for the asteroid tug. Also under study has been the possibility of taking advantage of the Yarkovsky Effect by changing the albedo of a threatening object, perhaps by coating it with coal dust. If done cleverly, the change in albedo could change the target's thermal properties so that the Yarkovsky Effect would move it out of a collision course with Earth. As with the other continuous techniques, this would require a long lead time to be effective, although in contrast with the other techniques, it would not require the continuous operation of a spacecraft.

Although not mitigation per se, there is also the option to do nothing and allow an impact to occur. This will be the response for some predicted impacts, for instance for those objects too small to have an appreciable effect. It might also be determined that an impact will occur in the ocean and the most cost-effective response will be evacuation of coastal areas rather than attempting to stop the impact.

Beyond the technical questions, there are also many policy and legal issues that potentially complicate any mitigation efforts. It is not clear whether an international body like the United Nations would claim jurisdiction over potential mitigation efforts, or if individual spacefaring countries would act independently or in consortia. From a financial liability standpoint, it is not clear whether the responsibility for an unsuccessful mitigation effort would be assigned to any party, and whether NASA or the

United States, for instance, could be sued. Along similar lines, thresholds above which action would be taken are also not established—if a mid-Pacific impact would devastate some sparsely populated islands, it is not clear whether the proper response would be to attempt mitigation or simply evacuate the area.

There is also no consensus for a threshold probability at which action will be taken. From a mitigation perspective, the earlier action can be taken the less force is required to succeed. On the other hand, the uncertainties in impacts are often large enough that the estimated probabilities of impact are only a few percent until several months ahead of time. If mitigation missions are launched each time the probability of impact reaches 2 percent, for argument's sake, 49 of 50 will have been unnecessary. Given the large cost of mitigation missions, and the risks present in the missions themselves, this presents a quandary.

SUMMARY

The Earth has been impacted by asteroids and comets uncounted times in its history, up to the present time. These impacts have been responsible for large, wide-ranging extinctions. Ongoing surveys of near-Earth objects have discovered nearly a thousand bodies large enough to have a drastic effect on human civilization if they impacted, though no objects on collision courses. New surveys will focus on smaller objects capable of regional devastation. Scientists and engineers have identified a number of techniques for possible prevention of impacts, though some are controversial. Impulsive techniques are the ones that are closest to readiness, and thought to be the most appropriate in cases of little warning. Continuous techniques provide more control over the outcome, though require technology that may not be ready for some time. Beyond technological issues, political and ethical questions surrounding NEO impact and mitigation remain largely unasked, let alone resolved.

WEB SITES

The issues of near-Earth objects and the impact hazard they pose are the subject of several compilations on the Internet. The Planetary Society has a set of popular-level links at this URL: http://www.planetary.org/explore/topics/near_earth_objects.

The Minor Planet Center is the central clearinghouse for collecting NEO discoveries and observations and calculating orbits. This Web site has information about these observations, including technical information about astronomers: http://www.cfa.harvard.edu/iau/NEO/TheNEOPage.html.

The Jet Propulsion Laboratory (JPL) is in charge of NASA's effort to track and characterize NEOs and has a wide array of information at its Web site: http://neo.jpl.nasa.gov.

Also at the JPL Web site is the text of a recent report to the United States Congress compiled by experts: http://neo.jpl.nasa.gov/neo/report2007.html.

This site has a different look at the impact hazard by scientist and artist Bill Hartmann; it details his attempt to make a painting of the Tunguska Event as scientifically accurate as possible: http://www.psi.edu/projects/siberia/siberia.html.

12

Spacecraft Missions

Only in the last half-century has humankind been able to visit other members of the solar system. And only in the last quarter-century have small bodies been explored close-up by spacecraft, with the first visits to dwarf planets planned for the coming decade. Spacecraft missions have revolutionized our knowledge of asteroids and comets, and planned missions will further our knowledge even more. In this chapter, we will look at the various types of spacecraft missions that exist and the instrumentation they have carried (or will carry), consider the unique benefits and limitations of spacecraft missions, and discuss what has been done and what we can expect in the future.

DELTA-V AND ORBITS

When considering the ease (or difficulty!) of reaching a particular target object, spacecraft navigational engineers must take the difference or similarity to the Earth's orbit into account. Because the Earth and the target body are both orbiting the Sun at different speeds, spacecraft must carry enough fuel to change from the Earth's orbit to the target's. The change in speed required for this orbital change is called **delta-v**, which is a standard measure of the relative difficulty of reaching an object. For the Moon, delta-v is roughly 6 km/s. Some near-Earth objects (NEOs) have delta-v as low as 4 km/s, meaning they require less fuel for a mission than do missions to the surface of the Moon. Orbital speeds are related to the inverse square root of the distance to the central body, so an object at 4 AU would have an orbital speed roughly half that of Earth, while one at 100 AU would be traveling roughly one-tenth as quickly as the Earth. This means that it is relatively easier in terms of fuel consumption to reach outer solar system bodies than you might think—in terms of delta-v, it is only about two to three times more costly to reach the distance of Pluto and even Eris than it is to reach Mars. In practice, however, delta-v requirements to visit small solar system bodies are larger than the best-case scenarios because unlike the major planets, their orbits are seldom near the same plane as the Earth. Changing orbit planes is very costly in terms of extra fuel required.

The simplest calculation of delta-v, still somewhat more involved than can be presented here, assumes a particular type of orbit called a Hohmann transfer orbit. Such an orbit is an ellipse tangent to both the Earth's and target's orbit. For a body like Mars, the transfer orbit's **periapse** is at the Earth's orbit and its **apoapse** is at Mars's orbit further from the Sun. For targets closer to the Sun than the Earth is, the transfer orbit's apoapse is at the Earth and its periapse is at the target.

The path followed by a spacecraft (also called its **trajectory**) is often more complicated than a Hohmann transfer orbit. It was recognized in the mid 1970s that close passes near planets can be used to deflect spacecraft trajectories and conserve fuel. Many missions now take advantage of these **gravity assists**, often using the Earth. Figure 12.1 shows a schematic of Dawn's trajectory as an example.

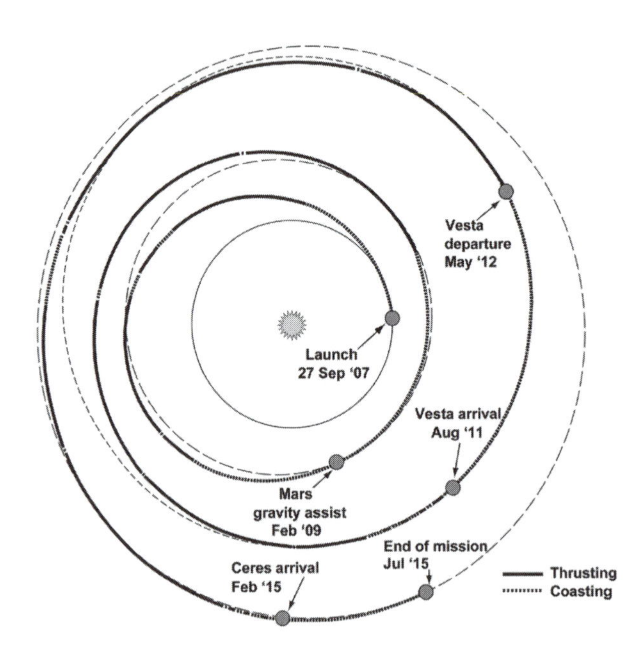

Figure 12.1 Dawn will visit both Vesta and Ceres, helped along by a close pass to Mars that will use that planet's gravity to alter its orbit. This diagram shows the critical dates in the mission timeline, along with a schematic view of the positions and orbits of Earth, Mars, Vesta, Ceres, and Dawn's path. NASA.

WHAT GOOD ARE MISSIONS?

Missions to small bodies (or indeed to any object) provide the opportunity to perform studies that are impossible via remote sensing. Instead of calculating a body's dimensions by measuring the brightness changes of a point of light (as described in Chapter 6), the size and shape of an asteroid or comet can be directly seen. Instead of getting a reflectance spectrum that measures the average composition, differences across an object can be measured. Furthermore, some data can be taken that is impossible to obtain from the Earth—the regolith of an asteroid or the coma and tail of a comet can be directly sampled, or the elemental ratios of a body can be measured. In the most complicated mission scenarios, material can be brought back to be measured in laboratories on Earth.

In addition to these purely scientific reasons for carrying out missions, there are other motivations that sometimes come into play. Missions can serve roles as engineering or technology demonstrations. For instance, the *Deep Space 1* mission to Comet Borrelly was largely intended as a test of a new type of propulsion, which has subsequently been used on other missions. The exploration of outer space is seen by many as lending prestige to a nation and as a means of stimulating interest in science and technology among its population. Missions have also been the source of international cooperation through the decades, as one nation will often participate in building instruments for another nation.

To date, there have been four leading players in space exploration beyond Earth orbit: the United States of America (through its space agency NASA), Japan (its space agency is called JAXA), the Soviet Union (before its dissolution in 1991), and the European Space Agency or ESA, a consortium of 17 European nations (though most of its funding comes from the United Kingdom, France, Germany, and Italy). Additional nations are beginning to have interest and capabilities in planetary exploration, with Russia expected to launch a mission to Phobos in the coming years and India and China both launching lunar missions and planning further exploration. In addition, ESA member states Germany and France have shown some interest in independently undertaking missions to visit small bodies.

MISSION STYLES

There are several different types of spacecraft missions, varying in complexity and difficulty, as well as in the amount of data return and the instrumentation typically carried. These mission types can generally be described as **flybys, rendezvous, landers**, and **sample returns**. The set of instruments carried on a mission is typically called the **payload.**

Table 12.1. Past, Present, and Future Missions to Asteroids, Comets, and Dwarf Planets

Object	Mission	Year	Type
Comet Giacobinni-Zinner	ICE	1985	flyby
Comet Halley	*Vega 1, 2; Giotto; Suisei*	1986	flyby
951 Gaspra	*Galileo*	1991	flyby
Comet Grigg-Skjellerup	*Giotto*	1992	flyby
243 Ida	*Galileo*	1993	flyby
253 Mathilde	*NEAR Shoemaker*	1997	flyby
9969 Braille	*Deep Space 1*	1999	flyby
433 Eros	*NEAR Shoemaker*	2000-2001	orbiter, lander
Comet Borrelly	*Deep Space 1*	2001	flyby
Comet Wild 2	*Stardust*	2004	flyby, sample return
Comet Tempel 1	*Deep Impact, Stardust* NExT	2005, 2011	flyby
25143 Itokawa	*Hayabusa*	2005-2007	orbiter, sample return
2867 Steins	*Rosetta*	2008	flyby
21 Lutetia	*Rosetta*	2010	flyby
Comet Hartley	*Deep Impact*/ EPOXI	2010	flyby
4 Vesta	*Dawn*	2011	orbiter
Comet Churyumov-Gerasimenko	*Rosetta*	2014-2015	orbiter, lander
134340 Pluto	*New Horizons*	2015	flyby
1 Ceres	*Dawn*	2015	orbiter

Flybys

The easiest mission type to undertake in many ways is the flyby. A flyby mission has an orbit that allows a close pass to the target, during which data are collected. The delta-v requirements can be quite modest, but the encounter, or duration of the data collection period, can be short. Flybys were the first planetary missions that were performed by spacefaring nations, and have been the means by which initial reconnaissance of objects have been undertaken. Historical examples are the *Voyager 1* and *2* spacecraft of the 1970s and 1980s, which conducted a series of flybys of Jupiter and Saturn, with *Voyager 2* adding flybys of Uranus and Neptune.

Similarly, the first views of the small bodies have been via flyby. The Soviet *Vega 1* mission obtained the first images of a cometary nucleus in 1986 as it encountered Comet Halley, as did several other international spacecraft, and the first spacecraft to visit an asteroid, *Galileo*, did so en route to Jupiter.

Because they spend only a limited time near their target body, flyby spacecraft typically have instruments suited for quick, remote observations. Instruments commonly on board flyby spacecraft (indeed on most spacecraft in general) include cameras for imaging and spectrometers to measure reflectance properties. Some will also carry a **LIDAR** (light detection and ranging), which measures the time between emitting and detecting a pulse from a laser in order to measure distances very precisely.

During Comet Halley's most recent visit to the inner solar system in 1985–1986, several spacecraft (collectively referred to as the "Halley Armada") performed flybys through its tail and coma. These allowed **mass spectrometers** to be used. Mass spectrometers ionize samples and then measure how those ions act in electromagnetic fields to deduce their atomic weight. As a result, the elemental abundances can be measured with some accuracy.

Finally, a close pass by an object can allow its mass to be measured. This is done by observing the spacecraft orbit before and after the closest approach and measuring the deflection caused by the object. However, the precision of the measured mass is a sensitive function of how small the close approach distance is, as well as how fast the flyby speed is. Slow speeds and close passes provide much better data than fast flybys and distant passes.

Asteroids are sufficiently numerous that there are often opportunities for spacecraft flybys while they are traveling to a different destination. As mentioned before, the first asteroid encounters were flybys, as the *Galileo* spacecraft passed the asteroids Gaspra and Ida en route to Jupiter. *NEAR Shoemaker* encountered the asteroid Mathilde on the way to a rendezvous with Eros. Similarly, the *Deep Space 1* mission had a flyby of the asteroid Braille on the way to Comet Borrelly. The *CONTOUR* mission was designed to fly by at least two cometary nuclei (with the possibility of a third), but was lost before its first encounter. More recently, the asteroid Steins was visited by the European *Rosetta* spacecraft in mid-2008, with a second flyby planned before its final destination is reached.

An interesting twist on a traditional flyby mission was taken by the *Deep Impact* spacecraft. Shortly before its encounter with Comet Tempel 1, it released a 370-kg probe with a camera designed to crash into the comet, while the main spacecraft flew by. The scientific goals were to see if cometary activity could be induced and allow crater formation to be observed on a large scale. At the same time, the comet was observed by a large number of astronomers on Earth as well as orbiting observatories. The mission was very successful, although ironically neither of the goals were met—a large amount of dust was created in the impact, which blocked a view of the crater formation, and an outburst was not started by the impact. However, the data collected and observations of the dust provided a much greater understanding of cometary surfaces (see Chapter 8).

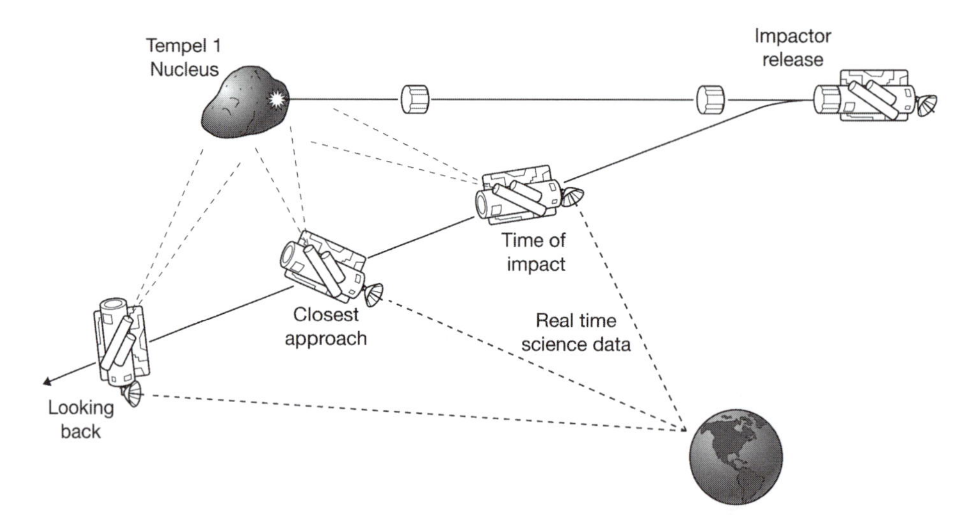

Figure 12.2 The Deep Impact mission released an impactor and flew by the target comet watching the impact. In this cartoon timeline, the spacecraft is moving from right to left. NASA.

The Halley Armada

The return of Comet Halley to the inner solar system in 1985 provided an unprecedented opportunity for spacefaring nations. Plans for missions to encounter Halley were in the works by NASA, the European Space Agency (ESA), the Soviet Union, and Japan by the late 1970s. At that point, there were hopes for a joint NASA/ESA mission to fly by Halley en route to a rendezvous with another comet a few years later. Budget pressures caused NASA to drop out of the joint project, leaving ESA to run a modified version of the mission alone. ESA named the mission *Giotto*, after a fourteenth-century artist who included a rendition of Comet Halley in his works. In an effort to find some way to study Halley, NASA rededicated the *International Sun/Earth Explorer 3* satellite from studying the Earth's magnetic field, sending it via a complicated series of maneuvers and gravity assists to the vicinity of Comet Halley and renaming the spacecraft the *International Cometary Explorer* (ICE)

In addition to *Giotto* and ICE, there were four other spacecraft that had flyby encounters with Comet Halley, a group collectively called the "Halley Armada": *Vega 1* and *Vega 2* (both Soviet/French missions), and *Suisei* and *Sakigake* (both Japanese). The encounters all occurred during March 1986, all but ICE during the period of March 6–14. The Halley Armada approach allowed an international division of labor, which increased the scientific return of each mission relative to its cost.

Ironically, while the Halley Armada was designed to be humanity's first encounter with a comet, ICE encountered Comet Giocobini-Zinner in September 1985 while en route to the vicinity of Comet Halley. It thus became the first spacecraft to encounter two small bodies. Other members of the Halley Armada also had additional targets: *Vega 1* and *2* both encountered Venus prior to Halley and dropped probes on that planet, while *Giotto* encountered Comet Grigg-Skjellerup in 1990.

RENDEZVOUS

With additional fuel and the proper timing, a spacecraft can rendezvous with its target rather than merely fly by. Smaller objects have weak enough gravity that a spacecraft will fly in formation with it around the Sun rather than orbit it. Larger objects such as Vesta have sufficiently strong gravity that a spacecraft will orbit it. In either case, however, a rendezvous mission spends a significantly longer time at its target than a flyby does.

Because of this, the amount and types of data that can be returned during a rendezvous mission are much larger than for a flyby. Images and spectra can be obtained over the entire surface of the target body instead of just the side visible during the flyby. Use of a LIDAR combined with knowledge of the spacecraft's position can provide a very accurate shape over an object's entire surface, allowing the calculation of a precise volume. Because rendezvous missions allow the determination of the target's mass much better than during flybys, this combination results in good values for the density for the target.

In addition, other instruments have been carried on small body rendezvous missions. Gamma-ray spectrometers (GRS) measure gamma rays (high-energy photons) emitted by elements when they have themselves been hit by high-energy cosmic rays. The various elements emit gamma rays with

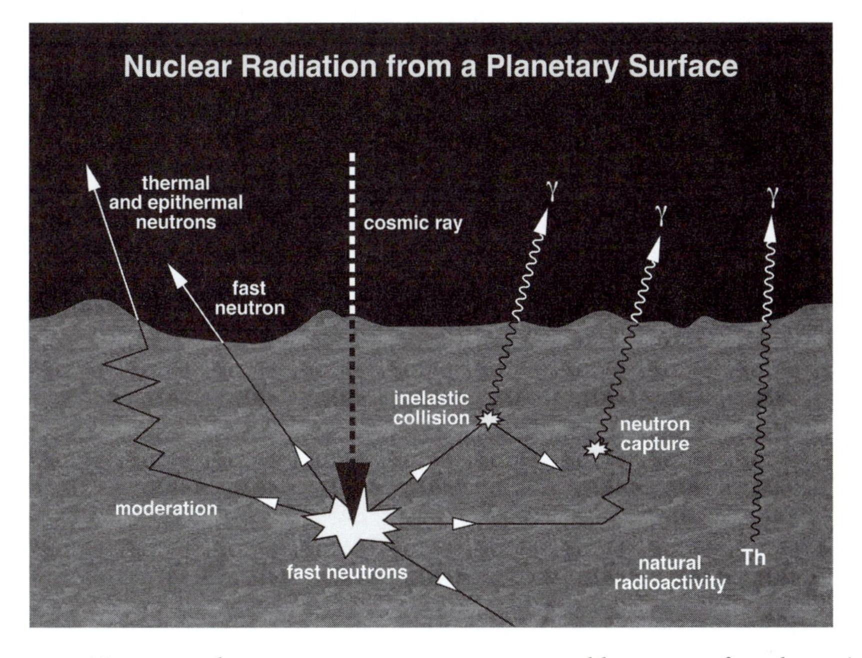

Figure 12.3 Neutron and gamma ray spectrometers are used by spacecraft to determine compositions. They take advantage of differences in the way elements react to high-energy photons to measure the relative concentration of those elements. They have been common instruments on planetary missions, including *NEAR Shoemaker* and Dawn. NASA/JPL.

different energies, allowing them to be distinguished from one another. The resulting data give the relative abundances of many important elements including carbon, oxygen, silicon, iron, and magnesium in the top meter or so of the target's crust. The rate of gamma ray emissions is relatively slow, however, and therefore a relatively long time is required to collect the necessary data. Flybys occur much too quickly, but *NEAR Shoemaker* and *Hayabusa* both carried GRS instruments, which showed that Eros and Itokawa were similar to ordinary chondrite meteorites in composition (see Chapter 7). X-ray spectrometers (XRS) work similarly, though the photons have different energies and originate from the Sun rather than cosmic rays. Often a neutron spectrometer (NS) is also included; these instruments measure the flux of neutrons with differing speeds from a surface, which allows the presence and abundance of hydrogen to be measured. The *Mars Odyssey* spacecraft used a NS to measure subsurface water ice on Mars. The *Dawn* spacecraft is carrying an instrument called the Gamma Ray and Neutron Detector (GRAND) to Vesta and Ceres. *Rosetta* will visit Comet Churyumov-Gerasimenko in 2014 and carries 11 instruments from Europe and the United States to provide comprehensive information about their target.

LANDERS

An additional step in complexity can be taken by landing on the target. Landing on objects like Mars or the Moon requires additional delta-v, but again the weak gravity of the small bodies means that it is not difficult to change the spacecraft's speed to be able to land. Ironically, however, it is the weak gravity that causes other difficulties. On Mars or the Moon, the gravity is strong enough that engineers can easily calculate the speed at which the spacecraft will approach the surface when in free fall, and the spacecraft can be programmed to fire retrorockets when it reaches a particular speed in order to give a soft landing. Any small uncertainties in the falling speed or imprecision in the rocket firing will be a tiny percentage of the total speed and thus won't have a big effect.

On small bodies, on the other hand, the gravity is small. Escape speed on Eros is 10 m/s, roughly the speed of an Olympic sprinter. On Itokawa, it is 20 cm/s, much slower even than a comfortable stroll. Landing speeds will be similar to these escape speeds. At these speeds, inaccuracies of only a meter or so in position, or a fraction of a second in firing a rocket, can spell the difference between a soft landing and a crash or a bounce. Indeed, the *Hayabusa* spacecraft suffered several slow-motion crashes into Itokawa during its mission. An additional complication is due to the irregular shape of small bodies, which results in variations of even this weak gravity depending on location. Finally, the weak gravity means that thought must be given to anchoring the spacecraft on the asteroidal or cometary surface, otherwise the lander might accidentally throw itself off the surface entirely!

Once on the surface, however, additional possibilities are available to landed spacecraft relative to rendezvous missions. The time required for collecting useful data with a GRS or XRS decreases with decreasing distance from the object, so landing on the surface is optimal. Cameras can provide extremely detailed images of the surface, though they must be designed to operate in focus at very short distances. Additional instruments can be used on the soil, either by placing the instrument directly on the surface or by scooping regolith and bringing it into an experiment chamber inside the spacecraft.

Although these techniques have not been used on a small body mission, they have been used on landers on Venus, the Moon, and Mars many times. They are also planned for the Philae lander, which is part of the *Rosetta* mission. Philae will harpoon itself to the surface of Comet Churyumov-Gerasimenko (often called "C-G") in 2014, and will also drill itself into the surface. The harpoons will have instruments of their own, measuring the density and strength properties of the surface. Philae's instruments will provide isotopic and mineralogical data about C-G's surface in much greater detail than is possible from a rendezvous alone. It will also drill 20 cm below the surface and study that subsurface material. As with many planetary landers, the *Rosetta* mission has a rendezvous craft in addition to a lander. This strategy allows a smaller lander, and the opportunity to use the rendezvous craft to relay communications to Earth, which allows the lander to only need a small antenna.

The first asteroid lander was *NEAR Shoemaker*, which was designed as a rendezvous mission. After its mission objectives were achieved it was brought in for a landing on a pond on Eros as an experiment, which brilliantly succeeded. *NEAR* was not designed as a lander, and its instruments

Figure 12.4 *NEAR Shoemaker* visited the asteroids Mathilde and Eros, carrying several instruments. AP Photo/NASA.

were downward-pointing, so the camera and spectrometer were not usable on the surface, but the GRS continued to operate and provided additional data while *NEAR* was landed. The order of operations performed by *NEAR*, with a rendezvous phase allowing detailed mapping and informing the choice of a landing site, is seen as a model for most future small body lander missions.

SAMPLE RETURN

No matter how sophisticated the instruments carried on a mission, they are inevitably less capable than the ones found in earthbound laboratories. This is because the amount of preparation and lead time for missions is sufficiently long that even the most modern equipment can become outdated by the time of launch. In addition, there are special requirements for equipment on spacecraft such that the most state-of-the-art instrumentation has typically not been cleared for flight. As a result, there is a great impetus to bring samples back from planetary bodies to Earth, where the most up-to-date laboratory equipment can be used for analysis.

The first sample return missions were the *Apollo* landings on the Moon. These returned nearly 400 kg of material from the Moon, and analysis of *Apollo* samples continues to this day. Humans are not likely to be on any small bodies in the coming decades, so any sample return would be robotic. There have been robotic sample returns from the Moon, with the Soviet *Luna 16*, *Luna 20*, and *Luna 24* missions bringing back roughly 300 g total. Sample returns have been proposed for Mars as well, perhaps occurring sometime in the 2020s.

Because they can contain material that has been largely untouched since the formation of the solar system, sample returns from asteroids and comets are of particular interest to scientists. The large amount of meteoritic material we have for study on Earth has been affected by the terrestrial environment, and so is not as pristine as something collected from an asteroidal surface. In addition, the collection of meteorites on Earth is biased; we know, for instance, that very weak material will not be able to survive the passage through the Earth's atmosphere that meteorites experience. This is true of the unconsolidated regolith of objects that produce meteorites as well as some parent bodies that may not be represented at all. This is also true of comets—we do not have any meteorites that scientists suspect are from comets, and therefore any samples must be retrieved via spacecraft. As a result, sample returns from the surfaces of a comet and an asteroid are ranked as high priorities by NASA and have been the subject of intense interest by European and Japanese scientists as well.

The *Apollo* and *Luna* sample returns both involved landers drilling into the lunar surface. The small body sample returns that have flown to date have used different approaches. The *Hayabusa* mission planners chose not to land on Itokawa because of the landing difficulties mentioned previously. Instead, they used a "touch and go" strategy, where sample collection would occur very quickly, during a short interval when part of *Hayabusa* was in contact with the asteroid surface. During that interval, *Hayabusa* was designed to fire a projectile at Itokawa's surface, knocking off material and collecting it for the ride back to Earth. Due to problems with the mission, it is still uncertain whether *Hayabusa* was able to collect any samples, although this question will be resolved upon *Hayabusa*'s return to Earth in 2010.

Due to the nature of comets, sample returns from these objects can be both easier and harder than for asteroids. For instance, the fact that a comet is constantly shedding grains of dust into space means that a landing is not technically required in order to collect samples. NASA's *Stardust* spacecraft took advantage of this in 2004 by flying through the coma of Comet Wild 2 carrying sample collectors made of **aerogel**. Aerogel is an exceedingly lightweight material specially made with a very low density (0.002 g/cm^3, over 1,000 times less dense than typical rocks). By using aerogel, scientists were able to slow down the coma particles from over 6 km/s, the relative speed of the comet and spacecraft, and capture the samples without destroying them. *Stardust* returned roughly a million dust particles, the vast majority much smaller than 100 μm in size. While analysis is ongoing and will continue for years, initial results suggest that Wild 2 contains grains that originated over a surprising range of distances in the solar system, with some possibly originating from outside the solar system entirely.

A sample from a cometary nucleus is even more desirable than a coma sample. The material found in the coma is only the silicate fraction of the comet's composition; the ices are absent after their sublimation. Better still,

Figure 12.5 The sample return mission Stardust used aerogel to slow down and capture dust from Comet Wild 2. Shown here is a closeup of the Stardust aerogel, with a track due to a captured dust grain visible just left of the center of the image. These grains are extracted one at a time and analyzed, with the analysis still under way. NASA.

from a scientist's point of view, would be a sample from beneath the surface of a comet, which would contain the least-processed material. This, however, would potentially run the risk of creating an outburst on the comet that could put the mission in danger. Regardless, sample returns from a comet nucleus have been identified as high priority by space programs around the world.

HUMAN EXPLORATION

While humans won't be standing on the surface of any small bodies in the next few decades, there have been proposals over the last 40 years to send astronauts (or cosmonauts) to visit asteroids. The benefits and drawbacks to human exploration are complicated, and there are strong feelings among both opponents and proponents. It is generally agreed that the *Apollo* astronauts were able to quickly identify interesting areas for exploration, and their ability to be flexible and improvise allowed problems to quickly be solved and on-site changes to be made. However, robotic missions are much cheaper and simpler than those that carry crews, and the failure of a robotic mission never has stakes or consequences as dire as the failure of a mission carrying people.

A piloted mission to an asteroid has been advocated as a natural stage in the exploration of Mars in particular and the solar system in general. This advocacy centers on the idea that asteroid missions can be intermediate in length between lunar missions and the flight time to Mars, allowing some systems to be tested in conditions that are not as demanding as a Martian mission. Over the long term, asteroids could be used for in-situ resource utilization (ISRU), where astronauts process asteroidal regolith to release water, oxygen, or metals that are useful for life support. In a similar vein, Phobos, the inner satellite of Mars and likely a captured asteroid, has been proposed as a base for the exploration of Mars.

A second rationale for human exploration of the small bodies has already appeared in movies—as a way to prevent the impact of a near-Earth object (NEO). In this case, some argue that the greater flexibility and problem-solving that humans showed on the Moon would be more than worth the extra expense, given the potential result of a large impact.

OBSERVATORIES IN SPACE

Although they are not missions *to* small bodies, orbiting satellite observatories have also played a role in small body studies. Observing from space removes the interference from the Earth's atmosphere, whether from the risk of clouds or bad weather, or the restricted wavelength range observable because of absorptions due to water, methane, and carbon dioxide in our atmosphere, or just the small-scale wind and turbulence that makes stars twinkle.

The most famous orbiting observatory is the Hubble Space Telescope (HST), which has observed asteroids, comets, and dwarf planets since its launch in the late 1980s. It has been used to discover satellites of Pluto, to make multiwavelength maps of the asteroid Vesta, and to observe the breakup and impact into Jupiter of Comet Shoemaker-Levy 9, among many other observations. Other observatories that cover other wavelengths have also played large roles in furthering small bodies research. The infrared spectral region can be quite difficult to observe from Earth, but is an important one for studying composition and can be used to determine the diameters of objects. Two spaceborne infrared observatories, the Infrared Astronomical Satellite (IRAS) and the Spitzer Space Telescope took very large amounts of asteroid data, both as purposeful targets and serendipitously (in much the same manner that astrophysicists in the early 1900s found their photographs of galaxies littered with previously unknown asteroids). The IRAS and Spitzer data continue to be used to this day.

There have been proposals to launch future observatories specifically to search for and catalog NEOs. With the ability to observe 24 hours/day and in all directions, such a satellite could more quickly find potentially hazardous objects than using only earthbound telescopes. Such proposals often suggest using telescopes sensitive to the infrared region, where NEOs are very bright relative to stars, rather than visible light. Studies have shown that to be most effective, such a telescope would orbit the Sun at Venus's distance so that objects interior to Earth's orbit could be more easily observed. While such observatories have not been approved, the first NEO-observing satellite, a Canadian effort that is designed to orbit the Earth, may launch early in the next decade.

LIMITATIONS OF SPACECRAFT MISSIONS

There are obviously a large number of ways that spacecraft missions to small bodies extend the capabilities of scientists and provide data that are difficult or impossible to obtain from the Earth. However, there are also drawbacks to spacecraft missions compared to other ways of studying the asteroids, comets, and dwarf planets.

The first and perhaps most obvious is cost. The missions that the United States has launched in the recent past to small bodies have had price tags of roughly 400–500 million U.S. dollars. The *New Horizons* mission to Pluto has cost even more. Budgetary information for non-U.S. missions is often not presented in the same fashion as U.S. missions, but building, launching, and operating spacecraft costs many times more than building and operating even the largest telescopes.

In addition, space missions are exceedingly complex undertakings. Telescopes and instruments on the Earth can be maintained relatively easily, occasionally as simply as by having someone with a screwdriver do a half-hour of work. Spacecraft, on the other hand, are required to function for years, with no hope of physical intervention. They must be able to work for long durations and designed to handle a wide range of conditions. Components that work perfectly on their own can sometimes give unexpected results when used together, and sometimes simple bad luck can intervene.

Because of the cost, there is pressure to ensure mission success, although truly ensuring mission success would make costs skyrocket. NASA in particular has at times emphasized cost concerns over reliability, preferring a steady stream of cheaper missions, some of which fail, to infrequent expensive missions with greater likelihood of success. However, NASA's policies have varied on this point through the years.

A final issue is time. From the time mission ideas are first considered until the time a mission is launched is typically several years. From launch until data are returned is often an additional few years. The *New Frontiers* mission to Pluto will spend nine years en route to its destination—it spent roughly twice that amount of time in development. Because of this, and the desire for reliability, the instruments and components on spacecraft are often older designs and have capabilities that are not nearly as advanced as what might be commonly available. For instance, the sizes of hard drives and processor speeds in on-board computers are usually nowhere near the capabilities found in relatively cheap personal computers. For example, the computing ability of the *Galileo* spacecraft, active until 2003, was similar to that of the first home video game systems of the 1970s. Again, this contrasts with the ability of ground-based observatories to quickly use new equipment and take advantage of new findings.

SUMMARY

While the foundation for our understanding of asteroids, comets, and dwarf planets comes from centuries of ground-based observations, spacecraft visits have provided great advances in our knowledge using techniques impossible to employ by any other means. Simple flybys provide close-up views of objects impossible to obtain from Earth, and can allow masses to be measured as well as the composition of minerals on the surface. More complex missions carry instruments that measure the relative abundances of elements in a body and can precisely measure shapes as well as provide global imagery, or even return pieces of the target to Earth for study in our laboratories.

With increased complexity come increased costs and an increased chance of mission failure. These possibilities also arise with increased travel time and distance from the Sun. While flybys of asteroids and comets have been accomplished, and *NEAR Shoemaker* has landed on the asteroid Eros (and an ESA cometary lander will operate in the next decade), the distance of the dwarf planets means that only flybys are being considered at this time, with the *New Horizons* mission to Pluto en route to a 2015 encounter.

As technology improves, difficult missions become easier and costs come down. In coming years we look forward to sample returns from comets and asteroids, flybys of additional objects, and further scientific breakthroughs from space missions.

WEB SITES

This is the Web site of Japan's *Hayabusa* mission (most current-day missions have a presence on the Internet): http://www.spacetoday.org/Japan/Japan/MUSES_C.html.

This is the Web site of Europe's *Rosetta* mission: http://www.esa.int/SPECIALS/Rosetta/index.html.

For those interested in taking part in the *Stardust* sample analysis, this Web site provides that opportunity, as well as information about the *Stardust* mission: http://stardustathome.ssl.berkeley.edu.

These Web sites offer popular-level descriptions and explanations of the gamma-ray and x-ray spectrometers. The MESSENGER mission to Mercury carries these two instruments, which are commonly found on small bodies missions: http://btc.montana.edu/messenger/instruments/grns.php and http://btc.montana.edu/messenger/instruments/xrs.php.

At this site, the user can access an online Hohmann Transfer calculator, with the ability to input various starting and ending locations: http://home.att.net/~ntdoug/smplhmn.html.

Missions of all kinds, from those that have flown to those that were more flights of fancy, are described at these blogs, including small bodies missions as well as those with other targets: http://robotexplorers.blogspot.com and http://beyondapollo.blogspot.com.

13

Interrelations

In the preceding chapters, we have considered the asteroids, comets, and dwarf planets from many different directions. We have looked at where they are found, and what they are made of. We have thought about how they formed and how they have evolved. In this final chapter, we will tackle the question of how much they form a single coherent grouping, and to what extent they are distinct from one another. We will also look at how they relate to the other non-planetary bodies in the solar system.

OBJECTS ON THE LINE

The distinctions between asteroids and comets are in some ways clear and in other ways quite blurry. As discussed in Chapter 1 and elsewhere, the original, observational distinction is that asteroids have "starlike" appearances in a telescope (under typical conditions), while comets are "fuzzy." Additional research in the centuries since these definitions were created has added a host of additional distinctions between typical comets and asteroids; as a general rule, asteroids are thought to be rocky, while comets have significant amounts of ice.

During the time of the solar nebula, planetesimals formed with a variety of compositions ranging from metal-rich and ice-free to mixtures of materials dominated by ice. The specific compositions were dependent upon temperature, and followed a trend called the condensation sequence, described in detail in Chapter 5. Looking at the distribution of spectral types of asteroids (discussed in Chapter 7) with solar distance, we can recognize a trend from an inner belt dominated by S-class asteroids through a middle belt mostly populated by C-class asteroids, to P asteroids in the outer belt. All these suggest a continuous trend in composition. The fact that different groups are more common in different parts of the belt shows that mixing between groups hasn't been a major factor, otherwise different areas of the asteroid belt would have similar fractions of each spectral class. If mixing hasn't been important, the obvious conclusion is that the different groups formed close to where they can be found today. The D-class spectra of Trojan asteroids seem to follow that trend, leading to the long-standing interpretation that they formed near 5 AU, where they are today. Those spectra of cometary nuclei that we currently have also have D-class spectra, while Centaurs and TNOs have a mix of spectral slopes along with evidence of water and methane ice on their surfaces.

However, with updated dynamical models over the past decade, astronomers have realized that the asteroids and comets may have formed over a variety of distances and moved from their formation locations. Indeed, the Trojan asteroids are now thought to have formed perhaps as far from the Sun as Uranus or Neptune and been perturbed into their current orbits. While we do not think this is true of most main-belt asteroids, it is possible that a few objects now in the asteroid belt might have formed elsewhere, or that objects formed in places where there are no small bodies today might have been luckily preserved far from their creation sites. As a result, we might expect to find objects with characteristics we associate both with comets and asteroids.

How could we recognize such objects? And how could we distinguish between theories that predict the Trojan asteroids formed at 5 AU from those that predict they formed at 15–30 AU? The most straightforward approach would be to obtain compositional information and compare it to what we expect. Organic compounds are more stable farther from the Sun than they are close to the Sun. When we compare the relative amounts of carbon and other elements (like silicon or magnesium) that we see, we find much more carbon in objects that formed in the outer solar system like Comet Halley than we find in meteorites that formed in the inner solar system. Although data are scarce, we might also expect the same to be true of nitrogen, which is relatively abundant in the outer solar system as nitrogen ice, ammonia (NH_3), and cyanide compounds (those with a CN triple bond). If objects from the outer solar system were scattered widely through the solar system, we might expect that various comets, outer solar system planetary satellites, and TNOs might have similar amounts of carbon and nitrogen, or that there would be no consistent connection between composition and solar distance. Obtaining the data necessary to make firm conclusions will likely take several dedicated space missions, and may lie decades in the future.

A study of bulk compositions at different solar distances could show whether objects have moved far since their formation was completed. A separate question is whether there was significant mixing very early on and if objects are composed of only locally available material. It is recognized that the asteroids and comets represent "leftovers" from planetary formation, representing the conditions at locations in the inner and outer solar system, respectively. In addition to the realization that objects formed at 30 AU may now be found at 5 AU, and vice versa, results from the *Stardust* mission to Comet Wild 2 suggest that at least in some cases objects may be composed of mineral grains formed over a wide range of distances. In fact, the silicate grains captured by *Stardust* are much more similar to asteroidal material than what was expected from a comet. This runs counter to established models of solar system formation that predict the material that accretes to form bodies is collected from a relatively narrow range of solar distances.

COMETS IN ASTEROIDS' CLOTHING

While most large-scale shuffling of orbits occurred billions of years ago, there are still individual objects today whose orbits are evolving, as touched upon in Chapter 3. Jupiter has been the most significant actor in perturbing and changing the orbits of small bodies throughout the history of the solar system. Close passes to Jupiter can make objects from the outer solar system into full-time residents of the inner solar system. However, by the laws of orbital mechanics, some characteristics of the original orbit can still be found in the new orbit.

As discussed in Chapter 3, an orbit can be described by a set of six orbital elements. A complex combination of three of them, the semi-major axis (or mean distance from the Sun), eccentricity (or how similar to a circle the orbit is), and inclination (or the angle between the plane of the orbit and the plane of the Earth's orbit), along with the semi-major axis of Jupiter is defined as the Tisserand parameter (T_j). The importance of the Tisserand parameter is that it stays nearly constant before and after interactions with Jupiter. For most comets, T_j is less than 3, while for most asteroids it is greater than 3. A Trojan asteroid of Jupiter in a perfectly circular orbit coplanar with Jupiter would have a T_j equal to 3. The Tisserand parameter is often used as a means of interpreting whether or not an object began in the outer solar system.

This is particularly useful for objects that share some characteristics with different groups of small bodies. One such group is the **Damocloids**, named after the object 5535 Damocles. Damocloids have orbits that look very much like long-period comets, but show no evidence of cometary activity. This has led to the interpretation that they are extinct comets, which once had tails and comae but have exhausted their store of near-surface volatiles. This would suggest that they are compositionally the same as comets even though they might appear to be asteroids.

While the origin of Damocloids may be easy to understand, and their orbits made an inner solar system origin obviously unlikely, the NEO population hosts a more difficult situation to untangle. As described in earlier chapters, bodies can evolve into near-Earth orbits from the asteroid belt or the outer solar system. Because comets are expected to have icy interiors compared to asteroids, and because a comet is expected to be weaker than asteroids of the same size, different strategies might be employed to deflect a threatening comet than a threatening asteroid. This makes the identification of extinct comets in the NEO population of more than scientific interest. It also adds to the desire to understand whether any objects with compositions intermediate between asteroids and comets exist and whether they can make their way into potentially hazardous orbits. The Tisserand parameter is used in combination with reflectance spectra to try and identify the most likely cometary candidates.

One of the best extinct comet candidates in the NEO population is 3200 Phaethon. Its orbit is very similar to that of the Geminid meteor shower, suggesting Phaethon is the Geminid parent body, where the material originates. All other known meteor shower parent bodies are comets, which has led to the interpretation that Phaethon is likely very similar to comets. However, other information about Phaethon points away from a cometary interpretation—it has never been seen to have any activity of any kind, and its Tisserand parameter value is much higher than 3. Its reflectance spectrum is similar to the C-class asteroids of the mid-asteroid belt rather than the steeply sloped spectra of outer-belt asteroids or cometary nuclei. Another favored candidate is 4015 Wilson-Harrington. This NEO was discovered in 1949 as a comet, and named Comet Wilson-Harrington after its discoverers. It was lost shortly after, but its orbit was connected with that of an asteroid discovered in 1979. It also has a C-class spectrum and a Tisserand parameter higher than 3. Wilson-Harrington has not shown any evidence of cometary activity since its original discovery, leaving astronomers puzzled about whether it is a comet that only occasionally has outbursts or if it was the victim of a collision in 1949 that threw off dust and ejecta and mimicked cometary activity.

"ACTIVATED ASTEROIDS" OR "MAIN-BELT COMETS"?

Even within the main asteroid belt the distinction between asteroids and comets can be blurred. Three objects have been found that orbit within the asteroid belt but occasionally have comae and tails like comets. Studies of their orbits show that these objects, called main-belt comets, must have been formed at their current locations near 3 AU rather than being formed in the outer solar system and transported to the inner solar system. For this reason, some astronomers prefer the term "activated asteroids" to "main-belt comets" since the latter could be interpreted as objects that started in the outer solar system. As with Wilson-Harrington, the first-known main-belt comet, 7968 Elst-Pizarro, was argued to possibly be the victim of a collision (with dust thrown off from the impact spoofing a cometary coma) until a second bout of cometary activity was seen several years after the first. The discovery of two additional main-belt comets (2005 U1 and 1999 RE70) made the collision hypothesis exceedingly unlikely—given the estimated impact rate in the asteroid belt, the odds that three objects on similar orbits would suffer impacts within such a very short time are prohibitively long.

All three main-belt comets are members of the Themis dynamical family, suggesting a common origin. Themis itself is much larger than the main-belt comets, and has a spectrum like C-class asteroids. Themis itself shows no sign of cometary activity. However, the current hypothesis is that all of the members of the Themis family likely have ice at relatively shallow depths, and that the ice became exposed on the main-belt comets. At this point, the exposed ice began to sublime and carry off dust, just as with more typical comets. Interestingly, reflectance spectra of Themis show some features that have been interpreted as due to water ice. Observations are still ongoing to try and determine how many of the more than 500 known Themis family objects show cometary activity. The existence of icy objects in the Themis family also raises the possibility that other unrelated objects in the outer asteroid belt could be ice-rich.

THE SMALL BODIES AND THE PLANETARY SATELLITES: PHOBOS AND DEIMOS

The satellites of Mars, Phobos and Deimos, were first imaged close-up by the *Mariner 9* spacecraft in the early 1970s. As small, irregularly shaped objects, they were immediately suspected to be captured asteroids, and in the decades before the *Galileo* spacecraft first encountered the asteroid Gaspra, knowledge about these objects was held to be true of asteroids in general. For instance, linear "grooves" are seen on Phobos and were interpreted as due to fractures created during impacts. Similar features have also been seen on asteroids like Gaspra and Eros (see Chapter 8), confirming this suspicion. Likewise, some very large craters exist on Phobos and Deimos, much larger than were once expected possible—Stickney crater on Phobos is 9 km in diameter, while Phobos itself is only 25 km across on average. There is evidence on Deimos for a crater roughly 10 km across, while that body is only roughly 12 km across on average. Here too, later observations of additional objects found that asteroids have craters with diameters roughly the size of their own diameters relatively commonly. However, in other ways Phobos and Deimos are not like typical asteroids. Their location in orbit around Mars leads to re-accumulation of ejecta that would escape if they were orbiting alone. This is because while they have very low gravities and impact debris can relatively easily escape Phobos and Deimos, the speed required to escape Mars's gravity is much higher. Typically the ejecta from impacts onto the satellites remains in orbit around Mars, where it eventually re-impacts either Phobos or Deimos. As a result, their regolith is much deeper than we would expect on objects of the same size in the asteroid belt. Furthermore, there is some evidence that ejecta from Deimos may be coating parts of Phobos, a type of situation impossible for the majority of asteroids that are solo objects.

The origin of the Martian satellites is still not well understood. As detailed dynamical models became available, it has been realized that it is exceedingly difficult to create a scenario where Phobos and Deimos are captured objects, either both at once or one at a time. Confusingly, the reflectance spectra of Phobos and Deimos are most like outer-belt asteroids or even Trojan asteroids, which are among the most difficult objects for Mars to capture. The dynamical problems facing capture scenarios have led to the suggestion that the Martian satellites were themselves once part of Mars, blown off of its surface in a giant impact similar to the one that created the Earth's Moon.

Given the uncertainty surrounding the origin and composition of Phobos and Deimos, their relationship to the asteroids and comets is as-yet uncertain, even after several spacecraft have obtained data during operations around Mars. A Soviet attempt to land on Phobos during the 1980s ended in failure, but a sample return from Phobos is planned by the Russian Space Agency. It is hoped that in the coming decade we will better understand the relationship between Phobos, Deimos, and other small bodies.

DWARF PLANETS AND BIG MOONS

Pluto, Eris, and Ceres are the first three objects to have been classified as dwarf planets. However, there are host of other objects that would also share that definition if not for the fact that they are orbiting other planets and thus disqualified. Indeed, some of the largest planetary satellites, like Ganymede and Titan, rival Mercury in size.

All the outer planets, as well as the Earth, have satellites in the diameter range of roughly 1,000–2,500 km. This size range also includes all of the currently known dwarf planets. This creates an opportunity to study similar-sized objects with different compositions across a very wide range of solar distances, from the rocky Moon and Io through Ceres' rock/ice mix to the icy satellites of Uranus and distant Eris. The study of comparative planetology seeks to gain insight into solar system bodies by comparing them to one another and pooling what we know about all of them rather than studying each object independently. For instance, the similar sizes and formation locations of Triton and Pluto have led many scientists to use knowledge gained in the *Voyager 2* visit to Triton and apply it to Pluto. The computer models that are used to learn about the interiors of Ganymede and Callisto have been used to study Ceres and large asteroids, in hopes of determining how many of them may have icy mantles beneath their crusts.

THE IRREGULAR SATELLITES OF GIANT PLANETS

As just discussed, the giant planets have a number of well-known, large satellites: Jupiter has Io, Europa, Ganymede and Callisto; Saturn has Titan and Enceladus; and Neptune has Triton. However, each of these planets, as well as Uranus, also has a large retinue of distant, small satellites. To contrast them with the larger objects, and in view of their usually nonspherical shapes, these are usually called **irregular satellites.**

Jupiter has nearly 50 irregular satellites outside the orbit of Callisto. These seem to fall into at least three groups, seen in Figure 13.1. Perhaps the most surprising characteristic of the outermost satellites is that they have orbital inclinations greater than 90 degrees, which means that they orbit Jupiter in a **retrograde** direction opposite from the majority of planetary satellites in the solar system. The irregular satellites are extremely far from Jupiter, the closest more than 5 times farther than Callisto (the outermost of the large satellites) and most over 10 times farther, taking more than two years to orbit Jupiter.

Like Phobos and Deimos, the irregular satellites of Jupiter are thought to be captured objects rather than having formed in place. In particular, the existence of the retrograde satellites seems to require capture, since all of the studies of satellite formation show that objects formed in place must revolve in a prograde direction. Dynamical studies show that it is much

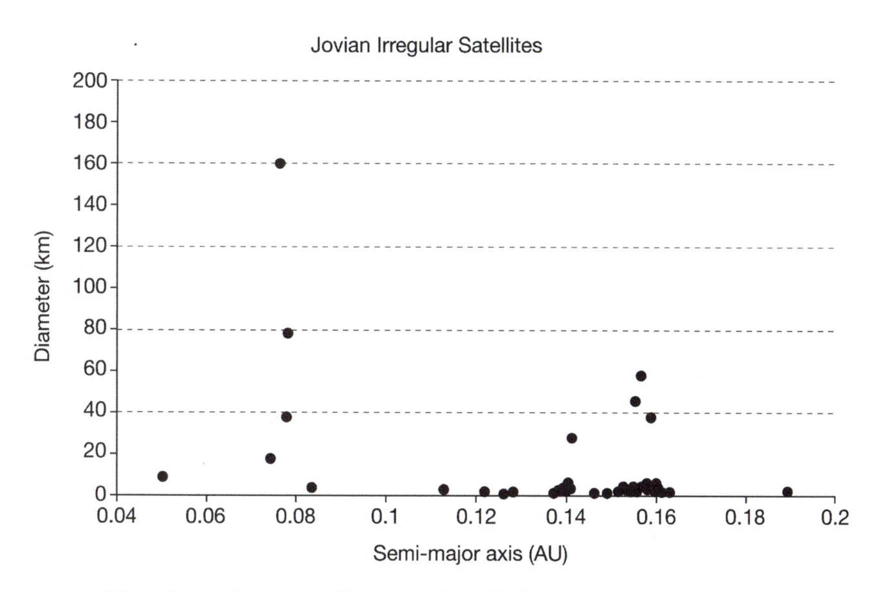

Figure 13.1 The giant planets well over a hundred irregular satellites among them, believed to be captured TNOs. The characteristics of those objects orbiting Jupiter are shown here, with each point representing a satellite. With few exceptions they are small (less than 20 km in diameter) and tens of millions of km from Jupiter, with some taking years to orbit that planet. For comparison with the plot, Mercury is 0.4 AU from the Sun. Illustration by Jeff Dixon.

easier for Jupiter to capture objects than Mars. The conditions that would allow capture early in solar system history (such as an extended atmosphere for Jupiter) no longer exist, although temporary captures are still happening today; Comet Shoemaker-Levy 9 was a satellite of Jupiter for perhaps 20 years before its impact in 1994. The simplest explanation for the large number of objects with similar orbits is that a few large objects were captured and then broke up either through impact or tidal stresses, resulting in the swarm of satellites seen around Jupiter today.

We know relatively little about these objects other than their orbits. Most of the irregular satellites are only a few kilometers in size, though a handful approach 50 km and the largest, Himalia, is nearly 200 km in diameter. Himalia was imaged by the *Cassini* and *New Horizons* spacecraft, but the images were taken from a great distance and show little detail other than an elongated shape. Detailed reflectance spectra are difficult to obtain because of the small sizes and faintness of the Jovian irregulars, but there appears to be at least two groupings, corresponding to the orbital groupings. The inner Himalia group, which have **prograde** orbits also have spectra like C-class asteroids. The outer retrograde group have spectra more like D-class asteroids and comets. There has been some suggestion that Himalia might have hydrated minerals on its surface, again like the carbonaceous chondrites, although that finding is yet to be confirmed by further study.

Figure 13.2 The best-observed irregular satellite is Phoebe. Cassini observed Phoebe in 2005, and scientists have concluded it is most likely a captured transneptunian object. It is also conjectured that most irregular satellites had such origins. NASA/NSSDC.

The irregular satellites of Saturn similarly fall into several groups, based on distance from Saturn and inclination. Again, these groups tend to share similar colors, some like C-class asteroids, and some D-class. Again, this is argued as evidence of an origin as captured TNOs. The spectral data has been bolstered by the *Cassini* spacecraft's close pass of the irregular satellite Phoebe. This pass obtained by far the best images of any irregular satellite of Saturn (see Figure 13.2), and also obtained a measurement of Phoebe's density of roughly 1,600 kg/m^3. This density suggests that Phoebe is roughly half rock and half ice, much rockier than the larger and closer Saturnian moons. This difference is further evidence that Phoebe did not form with the inner satellites but was captured, and by extension the same is true of the other irregular satellites.

Naming of Satellites

Until the mid-1800s, planetary satellites did not have a commonly accepted naming scheme. The Galilean satellites of Jupiter (Io, Europa, Ganymede, and Callisto) were formally called JI, JII, JIII, and JIV, their current names having been suggested soon after their discovery but never used in practice. John Herschel, son of Uranus's discoverer, William Herschel, suggested names for Saturn's satellites drawn from Greek mythology, while Shakespeare provided the inspiration for the satellites of Uranus.

Until the late 1900s, the number of planetary satellites was relatively small and few names were needed (all were drawn from Western literature or myths). Only 65 planetary satellites were known as the 1990s began (and 22 of those were discovered by the *Voyager* spacecraft), compared to the 166 known today. All these new discoveries were tiny irregular satellites.

After first trying to maintain the centuries-long precedent of naming satellites of Jupiter after love interests of that Roman god, the International Astronomical Union (IAU) allowed moons to be named for his descendents after discoveries outpaced his trysts. The irregular satellites of Saturn have been named after giants and monsters, with a Norse-inspired group, a Gallic-inspired group, and an Inuit-inspired group. Newly discovered Uranian satellites have continued to be named after characters in Shakespeare's work, while Neptune's moons are named for Greek sea deities.

Asteroidal and dwarf-planet satellites are named on a case-by-case basis, but usually the name has some connection with the parent-body name (as opposed to the case for Uranus, for instance). Occasionally, the satellite name has a whimsical aspect, as for instance with the 45 Eugenia system, which has one satellite named Petit-Prince after the *The Little Prince* by Saint-Exupéry and a more recently detected satellite unofficially called Petit-Princess.

Uranus and Neptune also have sets of irregular satellites, though little is known of most of them other than their orbits and estimates of their sizes. The exception is Neptune's moon Triton, the seventh-largest satellite in the entire solar system, with a diameter of 2,700 km (for comparison, the Earth's Moon is fifth largest, with a diameter of roughly 3,500 km). Triton orbits Neptune in a retrograde orbit, a sign that it was a captured object. Other satellites in the Neptune system, particularly Nereid, have orbits that show evidence of being perturbed when Triton was captured. The outermost satellites of Neptune have periods of more than 25 years, and also have retrograde orbits. Indeed, it is possible that the capture of Triton led to the loss of many satellites from Neptune, which alone among the giant planets is missing a set of several regular, similarly sized satellites.

CONNECTIONS TO OTHER STELLAR SYSTEMS

As mentioned previously, it is thought that 90–99 percent of the mass of TNOs and asteroids was ejected from the solar system. Therefore, in theory, material from our solar system is currently traveling through interstellar space and could find its way to other star systems. If the fraction of TNO loss theorized for our solar system is also true for other solar systems, there should be comets that formed around other stars occasionally making their way into our solar system. Indeed, by one estimate there should be 10^{23}–10^{24} such comets traveling through the galaxy (by comparison, there are 10^{11} stars in the galaxy).

Such interstellar comets should be easily detectable from their orbits. Objects bound to the Sun have elliptical orbits. Objects with very large semi-major axes but which come into the inner solar system, like long-period comets, have large orbital eccentricities (see Chapter 3), and the portions of the orbits we see look more like parabolas than ellipses. Objects unbound to the Sun, however, will be approaching with enough speed to escape from the Sun again (barring an exceptionally unlikely close pass to a planet), and their orbits will follow hyperbolic paths. Despite centuries of looking, no comet has ever been shown to have a hyperbolic orbit.

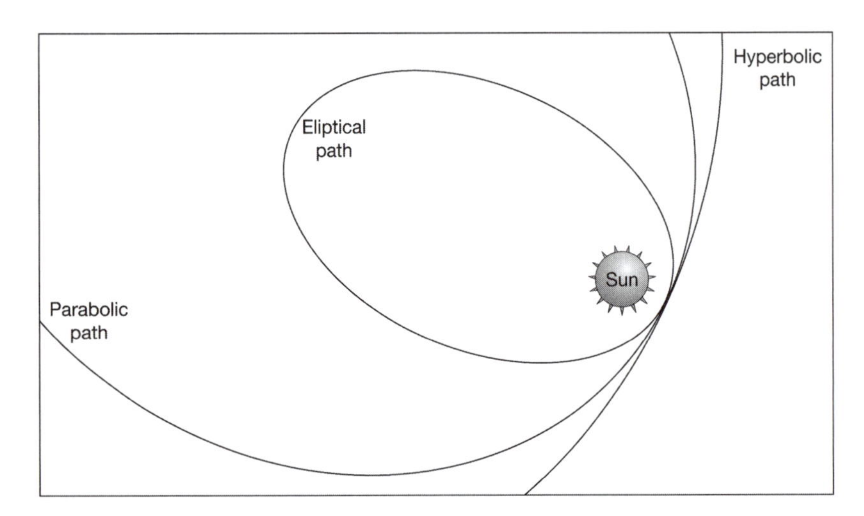

Figure 13.3 Orbits can have three shapes. Ellipses are the paths that all of the planets, asteroids, TNOs, and dwarf planets follow. Circles are a kind of ellipse. Parabolas are followed by long-period comets and indicate an object that is bound very weakly to the Sun. Hyperbolic paths relative to the Sun have never been seen for a comet, and would indicate an object visiting from outside the solar system. Illustration by Jeff Dixon.

Incredibly, however, astronomers are finding evidence of comets still orbiting other stars. While any objects in the Kuiper belts of other star systems are much too small to be seen from Earth, there is sufficient dust in those Kuiper belts to allow their extents to be detected. Our own Kuiper belt is thought to also have a similar dust ring, but ironically we are too close to detect it because its light is too spread out. However, sharp-eyed readers may have already seen the **zodiacal light**, a glow due to scattered light off of asteroidal dust visible near the horizon near sunrise or sunset.

While visits from interstellar comets have never been identified, there is very strong evidence for a stream of interstellar dust moving through our solar system. This dust was identified by the *Ulysses* spacecraft whose main mission is studying the Sun, with a secondary mission of studying interplanetary dust. However, it found some dust moving in a direction opposite that expected for interplanetary dust, and with a speed of 26 km/s, faster than the Sun's escape velocity (19–24 km/s over most of *Ulysses'* eccentric orbit) and interpreted as dust that originated outside of the solar system. Beyond a distance of 3 AU, most of the dust with sizes of about 1 μm is interstellar in origin rather than dust from our own solar system. The *Stardust* spacecraft sampled some of this interstellar dust and returned it to Earth in 2006, with analysis still underway.

SUMMARY

Comets and asteroids have long been seen as distinct objects, with comets considered ice-rich objects from the outer solar system and asteroids rocky, ice-poor objects from the inner solar system. However, it seems clear that some objects have elements of both groups. Scattering of objects throughout the solar system early in its history has potentially led to mixing, with objects that formed at many different solar distances found in relatively close proximity today. The satellites of the major planets also share some characteristics with the dwarf planets, asteroids, and comets, and in some cases they appear to be captured asteroids or transneptunian objects. The small bodies also have connections to other solar systems—scientists have observed evidence of Kuiper belts around other stars, and dust from interstellar space is streaming through our solar system and can be found in some primitive meteorites.

WEB SITES

This Web site has a trove of information describing the irregular satellites of the giant planets, put online by Scott Sheppard, who discovered many of these objects: http://www.dtm.ciw.edu/sheppard/satellites.

The official NASA Web page for Phoebe, satellite of Saturn and possible captured TNO: http://saturn.jpl.nasa.gov/science/moons/moonDetails.cfm?pageID=12.

At this Web site, Henry Hsieh, who discovered the main-belt comets (or activated asteroids), provides information about them, including links to more technical information: http://star.pst.qub.ac.uk/~hhh/mbcs.shtml.

Further information on the main-belt comets: http://www.planetary.org/blog/article/00000551.

A list of Damocloids and technical papers about objects that straddle the line between asteroids and comets is found at Dr. Yan Fernandez' Web page: http://www.physics.ucf.edu/~yfernandez/lowtj.html.

A discussion of the Geminid meteor shower and Phaethon, its parent body, were featured on the radio program *Earth and Sky* in 2007. This Web site has a transcript and audio file of the program: http://www.earthsky.org/radioshows/51876/geminid-meteor-storm-from-3200-phaethon.

Glossary

Absolute magnitude. The magnitude (measure of brightness) of an object calculated for a distance of 1.0 astronomical units from the observer and the Sun. It is related to the diameter of an object and often used as an estimate of size.
Absorption (light). One of the possible outcomes of an interaction between light and matter. Absorbed light adds energy to the material with which it interacts.
Accretion. The growth of a body via collision, where the colliding objects remain stuck together, resulting in an object with the combined mass of both.
Achondrite. A meteorite with no chondrules. The achondrites are one of the major classes of meteorites, and include objects that have experienced high heat and melting.
Activated asteroid. An object whose orbit is indistinguishable from a main-belt asteroid but shows cometary activity. Such objects are also called **main-belt comets.**
Aerogel. A very low-density material used to capture cometary and interplanetary dust by the *Stardust* sample return mission.
Albedo. The fraction of light reflected from a surface, varying from 0 (perfectly absorbing) to 1 (perfectly reflecting). Cometary nuclei and outer-belt asteroids typically have albedos near 0.04, with asteroid albedos commonly ranging up to 0.3–0.4.
Amor object. A near-Earth object (NEO) with an orbit that is always outside that of the Earth's.
Angle of repose. The greatest angle that a pile of loose material can support before slumping. Cliffs greater than this angle are evidence that a material has some cohesive strength.
Anhydrous mineral. A mineral lacking water in its structure.
Aphelion. The point in an object's orbit where it is furthest from the Sun.
Apoapse. The point in an object's orbit where it is furthest from its central body. The aphelion is the same as the apoapse for objects that orbit the Sun.
Apohele object. A near-Earth object (NEO) with an orbit that is always inside that of the Earth's.
Apollo object. A near-Earth object (NEO) with an orbit that crosses the Earth's but is on average outside of the Earth's.
Aqueous alteration. A type of metamorphism where chemical reactions in the presence of water change the composition and type of minerals present, often to include some minerals that incorporate water in their structure.

Areal mixture. A term describing a spectrum resulting from contributions from distinct, unmixed regions. In an areal mixture, a given photon only interacts with a single kind of material.

Astronomical unit (AU). The average distance from the Earth to the Sun, equal to roughly 150 million km. The AU is very commonly used as a convenient measure of distance in planetary studies.

Aten object. A near-Earth object (NEO) with an orbit that crosses the Earth's but is on average inside of the Earth's.

Band center. The central wavelength of an absorption or emission band in a reflectance or emission spectrum. The band center can be diagnostic of composition.

Band depth. The fraction of light absorbed at an absorption band center compared to what would be expected if the band were absent. The band depth can be diagnostic of the amount of a mineral that is present.

Blackbody radiation. Light given off by an object related to its temperature. A perfect blackbody absorbs all light that falls on it, and does not exist in nature. However, many objects act like blackbodies in some ways.

Blinking. A technique used to discover or characterize asteroids and comets, used in the discovery of Pluto. Two photographs or images are aligned, with each visible for a short period. Moving objects jump back and forth (or "blink") between frames, while stars and galaxies remain in place. Once done with rotating mirrors, blinking is now commonly done with software.

Bode's law. A mathematical progression found in the eighteenth century that successfully predicted the semi-major axes of Ceres and Uranus, but failed for Neptune. Today Bode's law is considered by most astronomers to have little predictive power.

Bolide. A particularly bright meteor.

Calcium-aluminum-rich inclusion (CAI). Calcium and aluminum-rich materials found in carbonaceous chondrite meteorites. These were the first solids formed in solar system history, and their study has provided insight into the formation of the solar system.

Carbonaceous chondrite. An important class of meteorites, often considered among the best-preserved relics of solar system formation. Some carbonaceous chondrites have been aqueously altered, and some include organic material.

Carbonates. A major mineral type on the Earth, containing CO_3 as part of its structure. Carbonates are also found in meteorites and have been found on Ceres.

Catastrophic collision. A collision destructive enough that the largest intact fragment after the collision is less than half of the mass of the original target.

C-complex asteroid. One of the major taxonomic divisions of asteroid spectra. C asteroids are the most common in the main belt as a whole and include Ceres, Pallas, and Mathilde. They are often associated with carbonaceous chondrite meteorites.

Centaur. A small body with an orbit between those of Jupiter and Neptune, often crossing the orbit of one of the giant planets. Centaurs are thought to have originated in the transneptunian region and have compositions similar to comets.

Chondrite. A meteorite that contains chondrules. Chondrites are one of the major classes of meteorites, and those meteorites most commonly seen to fall are chondrites. They have experienced relatively little heating and no melting since they formed.

Chondritic abundances. The relative amounts of elements found in chondrites, particularly carbonaceous chondrites. Chondritic abundances are similar to solar abundances when considering the elements found in rocks, although the Sun has much more of some other elements like hydrogen and helium. It is believed that the terrestrial planets have roughly chondritic abundances of the elements in bulk.
Chondrules. Small, glassy, round pieces of glass found in chondritic meteorites. These are thought to have formed via a short-lived heating event very early in solar system history.
Collisional families. *See* dynamical families.
Coma. The portion of a comet surrounding the nucleus, formed by and consisting of gas and dust being lifted off of the nucleus. Comets close to the Sun typically have their nuclei obscured by the coma, while at greater distances from the Sun the coma may be absent.
Cometary jets. Regions on cometary nuclei where enhanced ice sublimation leads to greater production of gas and dust in fan-like shapes, as well as the gas and dust in those fans.
Condensation sequence. The order in which a hot gas with solar abundances forms specific minerals as it cools down in ideal circumstances. The solar system is thought to have roughly followed such a sequence.
Continuum. In spectroscopy, the level of a spectrum upon which no absorptions or emissions are present. Band depths are measured relative to this continuum.
Crust. The outermost part of a differentiated object, usually formed by further processing of the mantle. The crust, to be stable, has a lower average density than the material below it.
Cryovulcanism. Low-temperature volcanic activity in which the magma is water. Cryovulcanism is found on outer planet satellites and expected on larger transneptunian objects.
Cybele group. A group of asteroids in a 3:2 mean-motion resonance with Jupiter.
Damocloid. An asteroidal object (that is, showing no cometary activity) found in a cometary orbit.
Daughter molecule. A molecule formed in a cometary coma through the breakup of other molecules through dissociation or other means.
D-class asteroid. One of the asteroid spectral classes, having steep spectral slopes in the visible region. D-class spectra are associated with the outer asteroid belt as well as cometary nuclei.
Delta-v. The amount of velocity change required for a spacecraft maneuver, or a set of spacecraft maneuvers, often used to describe the total velocity change required to get to a particular target.
D/H ratio. The ratio of deuterium (D) to hydrogen (H) found in a material.
Differentiation. The separation of an originally well-mixed object into layers, with the densest material at the center. This usually occurs after heating and melting, with separation into a core, mantle, and crust.
Dissociation. The separation of a molecule into constituent parts after interaction with an energetic photon or particle.
Dust tail. The trail of dust from a comet, swept from the coma by solar radiation forces. *See also* ion tail.
Dwarf planet. An object, according to the International Astronomical Union, that is in hydrostatic equilibrium but is not large enough to have "cleared its orbit."

Dynamical family. A group of objects that share similar orbits, suggesting that they originated on the same object, broken up by collision.
Dynamical lifetime. The average amount of time an object will exist before it can be expected to leave its current orbit.
Eccentricity. A measure of how circular an orbit is.
Ecliptic plane. The plane of the Earth's orbit around the Sun.
Ejecta. Debris created in a collision and thrown from the impact site.
Emission. Light given off by an object, either because of its temperature or as excited atoms return to their ground state.
Emission spectrum. The distribution of emitted light from an object, usually showing regions of increased emission (or lines) on top of a continuum.
Encounter. In space missions, the period over which a target body is subject to more-intense data collection.
Enstatite. The mineral name for magnesium-rich and iron-poor pyroxene. It is one of the dominant minerals found in enstatite chondrite and aubrite meteorites.
Enstatite chondrite. One of the major groups of chondrite meteorites. The enstatite chondrites are dominated by the mineral enstatite, and also contain iron-nickel metal. They are associated with the M-class asteroids and thought to have formed in the inner part of the asteroid belt.
Equilibrium condensation. A geochemical model of the early solar system that assumes solids condensed out of a nebula with the same composition as the Sun, with all chemical reactions going to completion.
Equilibrium saturation. The situation on a planetary, satellite, or small body surface where the addition of new impact craters will erase some older craters, leaving the overall number of craters on its surface roughly the same as before. Crater counting can only reveal surface ages until this state is reached, after which only a minimum age can be revealed.
Exobase. The level in an atmosphere above which atoms and molecules are more likely to escape into space than to collide with other atmospheric particles.
Extinct comet. An object with a cometary history and overall cometary composition that once, but no longer, shows cometary activity.
Fall. A meteorite whose Earth entry was observed and was collected shortly thereafter.
Find. A meteorite found without knowledge of its fall timing. Most finds occur in Antarctica via concerted searching.
Flyby. An encounter where the spacecraft is never in orbit around the target. Usually flybys are relatively short in duration.
Fragment species. *See* daughter molecule.
Frost line. The distance from the Sun at which water ice becomes stable on an airless surface or as a free-floating crystal. This term is usually used in relation to the early solar system rather than current times.
Gardening. Turnover and mixing of the top layers of regolith on an object due to impacts.
Gas tail. *See* ion tail.
Gravity assist. The use of a close planetary flyby to alter the path of a spacecraft. Gravity assists help reduce the amount of fuel required to make large orbital changes.

Grooves. Linear structures seen on Phobos, Ida, and some other asteroidal objects. While there are several interpretations for them, their exact formation mechanism is not agreed upon.

Halley family comet. A comet with an orbital period less than 200 years and a Tisserand parameter less than 2.

HED meteorite. A group of achondrite meteorites including the howardite, eucrite, and diogenite meteorite classes. The HED meteorites are thought to have originated on Vesta.

Hilda Group. A group of outer-belt asteroids in a 3:2 mean-motion resonance with Jupiter. These tend to be P- and D-class asteroids.

Hirayama family. *See* dynamical family.

Hungaria asteroids. Asteroids within 2.00 AU of the Sun, on the inner edge of the asteroid belt. Hungarias are associated with E-class asteroids.

Hydrated minerals. Minerals that have water as part of their structure. Some also include minerals with hydroxyl (OH) in their structure as hydrated minerals.

Hydrostatic equilibrium. The state of balance between the weight of overlying material and the ability of deeper material to support it. Solar system objects in this state are classified as planets or dwarf planets, depending upon additional dynamical factors.

Hypervelocity impact. An impact with a speed higher than the speed of sound in the target material. Most impacts in the solar system are hypervelocity impacts, which have results very different from those at slower speeds.

Igneous. A rock or mineral type formed through the cooling of magma or lava.

Immature. A regolith that has recently formed and has not been affected by processes such as solar-wind implantation or micrometeorite bombardment.

Inclination. A measure of the angle between the plane containing an object's orbit and the ecliptic plane.

Infrared (IR). A region of the electromagnetic spectrum stretching from roughly 750 nm (beyond the range of the human eye) to roughly 1 mm in wavelength.

In situ. Spacecraft investigations that occur on the surface of an object.

Interplanetary dust particle (IDP). Small grains derived from asteroidal collisions or blown off of comets, and collected in the Earth's atmosphere.

Intimate mixture. A mixture where different mineral components are in contact with one another. In an intimate mixture, a given photon has interacted with more than one kind of material.

Ion tail. A trail of ionized gas released from a comet, swept away by the solar wind.

Iron meteorite. One of the major groups of achondrite meteorites. Iron meteorites formed either as the cores of differentiated bodies or in large impact-created pools of molten rock.

Irregular satellite. One of the smaller, outer satellites of the giant planets. Most of the irregular satellites are also irregularly shaped, and many have unusual orbits. It has been speculated that they originated as captured asteroids or transneptunian objects.

Isothermal. Having the same temperature everywhere.

Isotopes. Atoms that have the same number of protons (thus belonging to the same element) but a different number of neutrons (and thus different masses).

Jupiter family comet. Comets with a Tisserand parameter greater than 2. Jupiter family comets have orbits that interact with Jupiter.

Kirkwood gaps. Regions in the asteroid belt that have been cleared of objects due to mean-motion resonances with Jupiter.
K-T boundary. A layer in Earth rocks that separates material deposited in the Cretaceous Period from the Tertiary Period. It is believed a large impact occurred in this time, with a mass extinction occurring as a result.
Kuiper belt object (KBO). A transneptunian object found in a low-inclination orbit between 30 and roughly 55 AU.
Lag deposit. A surface layer on a comet that has become devoid of volatiles due to their sublimation. A deep lag deposit can prevent further sublimation from deeper layers.
Lagrangian points. Places in a two-body system where a third, small body can also have a stable orbit.
Lander. A spacecraft that lands on an object, often conducting in situ studies.
LIDAR. Light Detection and Ranging, a technique used on spacecraft to measure the distance to and shape of planets and small bodies.
Lightcurve. The change in brightness with time of an object. For asteroids this is usually due to shape, though it could also include albedo effects.
Linear mixture. *See* areal mixture.
Lithophile. An element that "prefers" to be in rock rather than metal, and is left behind in a rocky mantle when a metallic core is formed in an object.
Long-period comet. A comet with a period longer than 200 years.
Macroporosity. The amount of porosity in an object due to cracks and void space.
Magmatic iron. A type of iron meteorite formed during the differentiation of an object, and once residing in a now-disrupted iron-nickel core.
Magnitude. A measure of the brightness of an object. Brighter objects have numerically smaller magnitudes.
Main asteroid belt. A region between roughly 2 and 3.5 AU containing most asteroids. In general, orbits in the main belt are stable over billions of years.
Main-belt comet. *See* activated asteroid.
Mantle. A layer in a differentiated body that is less dense than the core. In the inner solar system, the mantle is usually rocky over a metallic core, while in the outer solar system it is typically icy over a rocky/metallic core.
Mars crossers. Asteroids whose orbits cross those of Mars but are not near-Earth objects. These objects are not on stable orbits, and will become NEOs and be removed from the solar system via impact on relatively short time spans.
Mass spectrometer. An instrument often carried on spacecraft missions that measures the different atomic masses of particles, allowing elemental abundances to be determined.
Mass wasting. The large-scale movement of material by gravity, such as landslides.
Mature. Regolith that has been subjected to high amounts of processes such as solar wind implantation, ultraviolet light, and micrometeorite impacts. Mature regolith can have properties quite different from immature regolith and rocks of the same composition.
Mean motion resonance. An orbit with a period that is an exact fraction of the period of a large planet. The smaller the numbers in a resonance, the stronger it is. For instance, the 2:1 resonance (orbit period half that of a planet) is stronger than the 13:7 resonance (orbit period 7/13 that of a planet). Kirkwood gaps are at locations of mean motion resonances.
Mesosiderite. An achondrite meteorite group containing both metals and silicates.

Metamorphic. A rock type that has been subjected to heat, water, and/or pressure sufficient to alter minerals and textures but not sufficient for melting.
Meteorites. A piece of an extraterrestrial object, usually an asteroid, that has survived a fall through the Earth's atmosphere and been recovered on the ground.
Meteoroids. Objects that give rise to meteorites, as called before they enter the Earth's atmosphere. Often they are considered to be small (up to 50 m), otherwise they are considered asteroids.
Meteors. The flash of light and trail associated with dust and small objects burning up in the atmosphere.
Meteor shower. A period of high rates of meteors, seeming to originate from the same point on the sky, caused when the Earth's orbit intersects that of a dust stream. Meteor showers recur on annual timescales.
Meteor storm. A meteor shower with an extremely high rate of meteor activity.
Miller-Urey experiment. An experiment in the 1950s that simulated the conditions on the early Earth and demonstrated that organic materials could be created from inorganic starting materials. This has also been argued as similar to the origin of organic materials in meteorites.
Minerals. Collections of atoms in particular crystal structures.
Minor planets. A general term for asteroids and comets no longer in common use.
Mitigation. The prevention of asteroid or comet impacts into the Earth either by diversion or destruction of the impactor.
Monolith. An asteroid or comet that is a single, unfractured piece.
Mutual event. A period when a satellite moves in front of or behind its primary from the line of sight of an observer.
Near-Earth object (NEO). An object whose orbit comes within 0.3 AU of the Earth's orbit. NEOs are further subdivided into Aten, Amor, Apollo, and Apohele depending on the specifics of their orbit.
Noble gas. One of several elements that do not form compounds, including helium, neon, and argon.
Non-gravitational force. A force that alters a cometary or asteroidal orbit such that it cannot be modeled as simply under the influence of gravity. Typical nongravitational forces include forces due to the Yarkovsky effect, and due to gas and dust loss on a comet.
Nonlinear mixture. *See* intimate mixture.
Non-magmatic iron. An iron meteorite that did not originate in the core of an object.
Nucleus. The central, solid portion of a comet. Nuclei are often obscured by the cometary coma when close to the Sun.
Oblate. An object with a nearly spherical shape, but with a polar radius smaller than the equatorial radius.
Obliquity. The angle between the equatorial plane of an object and the plane of its orbit.
Occultation. The blocking of one object by another. Asteroidal occultations, where an asteroid moves in front of a star, have been used to determine shapes and sizes.
Olivine. A common mineral in terrestrial planets, small bodies, and meteorites, containing iron, magnesium, silicon, and oxygen.
Oort cloud. A region up to 50,000 AU from the Sun hypothesized to be the source of long-period comets.

Opposition. In astronomy, the point at which the Sun, Earth, and an object form a straight line with the Earth between the Sun and object, when viewed from high above the Earth's north pole. It is also the point where the object is highest in the sky at midnight when viewed from the Earth.

Optical depth. A measure of the amount of light scattered or absorbed by a coma or atmosphere, which allows an estimate of the amount and kind of material present.

Orbit. The path taken by one body around another, dominated by the force of gravity.

Orbital elements. The set of six values needed to completely describe an orbit. The standard ("Keplerian") orbital elements are called semi-major axis, eccentricity, inclination, mean anomaly, argument of periapse, and longitude of ascending node.

Orbital period. The amount of time required for an object to complete an orbit.

Ordinary chondrite. One of the major meteorite groups. Ordinary chondrites are by far the most common falling meteorites. They have compositions of olivine, pyroxene, and metal and remain relatively unaltered since they formed. They are associated with the S-complex asteroids.

Organic material. Compounds containing both carbon and hydrogen, usually restricted to solid material. It is found in carbonaceous chondrite meteorites and on cometary surfaces, and is expected on some asteroidal and dwarf planet surfaces as well.

Oxides. Minerals containing oxygen but not silicon.

Palermo Scale. A quantitative measure of the hazard from a given NEO, taking into account the time until a possible impact and the probability of impact. The Palermo Scale is intended for specialist rather than public use.

Pallasite. One of a class of mesosiderite meteorites, known for having large crystals of olivine mixed with metal.

Parallax. The change in relative position of objects caused by a change in viewing geometry. Parallax has been used for centuries to determine the distance to objects of interest.

Parent body. The object on which a meteorite originated, or the original object from which a dynamical family was formed.

Parent molecule. A molecule that is broken up via dissociation or other means to form daughter molecules.

Payload. The portion of a spacecraft that conducts scientific investigations.

Periapse. In an orbit, the closest point to the primary.

Perihelion. In an orbit, the closest point to the Sun.

Photometry. The study of brightness of astronomical objects.

Photons. Fundamental, massless particles of light.

Phyllosilicate. A class of minerals that are usually hydrous and formed via aqueous alteration.

Planetary embryos. Objects with diameters of roughly 100–1,000 km that are thought to have been common early in solar system history. Some planetary embryos may still survive as dwarf planets.

Planetesimal. Objects with diameters of roughly 1 km that accreted early in solar system history to form larger objects, including the planets.

Plutino. An object in a Pluto-like orbit, in the 2:3 mean-motion resonance with Neptune.

Plutoid. Defined by the International Astronomical Union (IAU) for objects that are both dwarf planets and transneptunian objects.
Point source. An object that is too small to be seen as anything other than a point of light in a telescope.
Ponds. Relatively flat, smooth, areas with fine-grained regolith on the surfaces of Eros and Itokawa, and hypothesized to be present on other objects.
Porosity. The fraction of a rock, or an object, which is occupied by void space.
Potentially hazardous asteroid (PHA). An object with a size of at least 150 m that passes within 0.05 AU of the Earth's orbit.
Poynting-Robertson drag. A nongravitational force on dust grains that causes them to spiral in toward the Sun.
Presolar grain. A dust grain that formed before the solar system formed. Presolar grains have been found in some chondrites.
Pre-stellar Nebula. *See* protoplanetary nebula.
Primary. The most massive body in a system, around which other objects are considered to orbit. The Earth is the Moon's primary, and the Sun is the Earth's primary.
Primitive achondrite. A group of meteorites that have experienced very limited melting, placing them intermediate between the chondrites and achondrites in many of their properties.
Production function. The distribution of impactor amounts and sizes for a given object, and how that distribution changes with time. Knowledge of the production function is required to allow surface ages to be calculated from crater counts.
Prograde. Rotation or revolution in a counterclockwise direction when viewed from above the north pole of an object or a system.
Protoplanetary nebula. A cloud of gas and dust from which planets and small bodies form after a period of collapse and accretion.
Pyroxene. A common mineral found in terrestrial planets, small bodies, and meteorites. It contains silicon, oxygen, iron, magnesium, and calcium, in varying amounts.
Radiant. The point in the sky from which meteor showers appear to originate, and from which they are named.
Radiation pressure. A nongravitational force on dust in the solar system, which tends to sweep them away from the Sun.
Radiogenic heat. Heat generated from the decay of radioactive elements.
Radiometric ages. Ages calculated from the relative amounts of radioactive elements and their decay products.
Radionuclides. Radioactive nuclei of atoms.
Reflection (light). A change in the path of light at the surface of a material without the light penetrating the material.
Refractory. A material that is stable at high temperatures and is expected to be one of the first ones formed in solar system history.
Regolith. Loose, broken-up material found at the surface of an object. On airless bodies, regolith is generated via collisions.
Regolith breccia. A meteorite type composed of regolith that has been hardened into a rock. Regolith breccias give us much of the information we have about asteroidal surfaces.

Remote sensing. Data collection on an object using observations that are done while not in contact with that object. Examples include imaging from orbit or ground-based spectroscopy.
Rendezvous. A mission type in which the spacecraft expends delta-v to end up in orbit around its target, or if its target is small and has weak gravity, in orbit around the Sun in close proximity to its target.
Resolved. An object that is large enough to be detected as more than a point source.
Resonances. Regions in the solar system where orbits are not stable over long periods of time due to the periodic gravitational attraction of the planets. Objects in resonances have their orbits modified by the planetary attraction, and usually end up impacting one of the planets or the Sun, or being thrown entirely out of the solar system.
Retrograde. Rotation or revolution in a clockwise direction when viewed from above the north pole of an object or a system.
Rubble pile. An object consisting of pieces of many sizes that are in contact with one another but are not bound by any forces other than gravity and friction. Rubble piles have relatively low density compared to the material they are composed of.
Runaway growth. A period during the formation of the solar system when it is thought that larger planetesimals grew at a much larger rate than smaller planetesimals. This is hypothesized to have resulted in the formation of a relatively small number of planetary embryos.
Sample return. A mission style where the spacecraft collects material from the target (either by landing or flying through a cloud of material) and returns it for analysis on Earth.
Scattered-disk object. A transneptunian object that has an eccentricity and/or inclination too high to place it in the Kuiper belt. It is thought that the scattered-disk objects have had encounters with Neptune.
S-complex asteroid. One of the major spectral classes for asteroids. The S complex includes most of the asteroids visited by spacecraft (Eros, Ida, Gaspra, Itokawa), and most of the known NEOs. S asteroids dominate the inner asteroid belt and are associated with ordinary chondrites and mesosiderites.
Secondary. In a binary or multiple system, those objects that are not the most massive ones. The Moon is the Earth's secondary. The Earth is one of a large number of secondaries orbiting the Sun.
Secular resonance. A region within the asteroid belt where the rate of precession of certain orbital elements is synchronized with those of the giant planets. This leads to rapid alteration of the asteroid orbits and removal of those objects from the asteroid belt.
Sedimentary. A rock type composed of material deposited in layers before being incorporated into the rock.
Seeing. A measure of the stillness of the atmosphere in terms of the angular resolution achievable with an arbitrarily sized telescope.
Semi-major axis. The mean orbital distance between an object and its primary.
Shattered. An object that has been thoroughly fractured, but whose pieces have not subsequently moved relative to one another.
Short-period comet. A comet with an orbital period less than 200 years.
Siderophile. An element that "prefers" to be in metal rather than rock, and moves into a metallic core when it is formed in an object.

Silicates. A class of minerals that have silicon and oxygen. Silicates are the most common minerals in the crust and mantle of the Earth, and are exceedingly common in meteorites and throughout the inner solar system and asteroid belt.
Size-frequency distribution. A description of how the amount of something (usually craters or objects) changes depending upon their diameter.
Small solar system body (SSSB). Defined by the IAU as the set of solar system objects that are neither planets nor dwarf planets.
Snow line. *See* frost line.
Solar abundances. The relative amounts of elements found in the Sun. Chondritic abundances are similar to solar abundances when considering the elements found in rocks, although the Sun has much more of some other elements like hydrogen and helium.
Solar nebula. A region of gas and dust that according to current thought gave rise to the Sun and planets after collapse and accretion.
Solid state greenhouse effect. A physical process in deep ice whereby light penetrates but the resulting heat cannot escape. This is thought to give rise to convection on some icy satellites and, in principle, TNOs, with surface features and cryovulcanism possible results.
Space weathering. A general term for the processes that mature and alter regolith properties, especially those that alter spectral properties as well.
Spectroscopy. The scientific field of studying the light from an object by separating it into its constituent wavelengths. Usually spectroscopy is used to determine an object's composition.
Spectrum. The distribution of light at different wavelengths from an object. Also sometimes more generally used to describe other distributions.
Sporadic meteor. A meteor not associated with a shower.
Stone. A meteorite that is neither stony-iron nor iron, including both chondrites and achondrites.
Stony-iron. A meteorite group that includes pallasites and mesosiderites, with both stony and metallic components.
Strength. The ability of a material to resist deformation or breakage.
Sublimation. The transitioning of a material (usually ice) from solid directly to vapor form without passing through a liquid stage. This occurs at low pressures and in a vacuum.
Synodic period. The time between oppositions of a body, or more generally, the time between repetitions of the same geometry between a body, the Earth, and the Sun.
Tail. Gas and dust released from a cometary nucleus and swept out of the coma by solar wind and radiation forces.
Terminal lunar cataclysm. A hypothesized period of increased collisions on the Moon roughly 4 billion years ago.
Terminator. The dividing line on an object between sunlit areas and areas in darkness.
Thermal inertia. A quantitative measure of the ease or difficulty of heating and cooling a particular substance. It is dependent upon several factors, including particle size and composition.
Thermal metamorphism. A type of metamorphism where chemical reactions driven by heat change the composition and type of minerals present, although the heat is not sufficient for melting.

Tisserand parameter. A measure of the influence of Jupiter upon an object's orbit, which also allows some insight into previous orbits an object may have had. NEOs with Tisserand parameter values less than 3 are considered possible comets.
Titius-Bode relation. *See* Bode's law.
Torino Scale. A qualitative measure of the hazard from a given NEO, based on its size and the current knowledge of its orbit. The Torino Scale is intended for public rather than scientific use.
Trajectory. The path followed by a spacecraft.
Transmitted light. In spectroscopy, light that has passed through a material.
Transneptunian objects (TNOs). Objects with orbits beyond that of Neptune. Often the TNO group is considered to include the Neptune Trojan objects, but it is not generally considered to include objects in the Oort cloud.
Trojan asteroid. An asteroid that orbits at L4 or L5 of a planet. Currently Jupiter, Mars, and Neptune are all known to have Trojan objects.
T-Tauri stage. A phase early in the life of a star where powerful stellar winds remove any remaining gas and dust from its nebula. This is generally considered the end of star and planet formation.
Ultraviolet (UV). A region of the electromagnetic spectrum stretching from roughly 400 nm (beyond the range of the human eye) to roughly 10 nm in wavelength.
Undifferentiated. An object that has not differentiated. Undifferentiated material has components with very different densities in close proximity to one another.
Unequilibrated. Material that contains minerals that formed under very different temperature and pressure conditions, indicating that the material as a whole has not come into equilibrium with one another.
Unresolved. *See* point source.
V-class asteroid. A member of an asteroid spectral class that has spectral properties very similar to the asteroid Vesta. V-class asteroids are associated with the HED meteorites.
Visible light. Light with wavelengths between roughly 400 and 750 nm. Visible light can be seen by the human eye, with differing colors ranging from violet (near 400 nm) through blue, green, orange, and yellow to red (near 750 nm).
Volatile. Material with a low boiling point, usually absent from inner solar system bodies or present only in planetary atmospheres. Volatile material is the last to condense in equilibrium condensation models.
Wavelength. The distance between successive crests or troughs in a wave. When considering light, which has wave-like and non-wave-like properties, the wavelength may be most easily considered to be a property related to energy.
X-complex asteroid. One of the major groupings in asteroid spectroscopy. The X class includes asteroids with widely differing albedos and compositions. It is associated with aubrite meteorites, enstatite chondrites, iron meteorites, and some carbonaceous chondrites, among others.
Yarkovsky Effect. A nongravitational force related to asymmetric absorption and reemission of light, altering an object's orbit. This force is only non-negligible for rocky objects of roughly 10–100 m in diameter, and even on those objects it can take millions of years to make large-scale changes.
Zodiacal light. A glow in the pre-dawn or post-dusk sky due to sunlight scattering off of the zodiacal cloud; dust generated from impacts in the asteroid belt and from comets.

Annotated Bibliography

BOOKS

Beatty, J. Kelly, Carolyn Collins Petersen, and Andrew L. Chaikin, eds. *The New Solar System.* 4th ed. Cambridge: Cambridge University Press, 1998.

An excellent collection of articles about planetary science topics, including asteroids, comets, and transneptunian objects. It was published before the IAU creation of the dwarf planet category.

Bell, Jim, and Jacqueline Mitton, eds. *Asteroid Rendezvous: NEAR Shoemaker's Adventures at Eros.* Cambridge: Cambridge University Press, 2002.

This volume includes contributions from the scientists involved in the mission describing its conception and its findings.

Binzel, Richard P., series ed. University of Arizona Space Science Series. Tucson: University of Arizona Press. 1979.

The University of Arizona Press's long-running Space Science Series publishes cutting-edge research on planetary sciences. This series is relatively technical, aimed at graduate school-level students and designed as a general reference for professionals. The series currently has 30 volumes, but the books in the series relevant to the topics discussed in this volume are listed below. The series is periodically updated, so, for example, *Asteroids III* supersedes *Asteroids II*; however, earlier versions of books can still provide interesting and useful information and therefore are also included in the following list.

Asteroids III. Bottke, William F., Paolo Paolicchi, Richard P. Binzel, and Alberto Cellino. 2002.

Asteroids II. Matthews, Mildred Shapley, Richard P. Binzel, and T. Gehrels. 1989.

Asteroids. Gehrels, T. 1979.

Comets II. Festou, Michel C., H. Uwe Keller, and Harold A. Weaver. 2004.

Comets. Wilkening, Laurel L., with the assistance of Mildred Shapley Matthews. 1982.

Hazards Due to Comets and Asteroids. Gehrels, T. 1995.

Meteorites and the Early Solar System II. Lauretta, Dante S., and Harry Y. McSween. 2006.

Meteorites and the Early Solar System. Kerridge, John F., and Mildred Shapley Matthews. 1988.

Pluto and Charon. Stern, S. Alan, and David J. Tholen. 1998.

Protostars and Planets V. Reipurth, Bo, David Jewett, and Klaus Keil. 2007. This volume is listed because it discusses solar system formation and includes chapters touching on small bodies topics.

The Solar System Beyond Neptune. Barucci, M. A., H. Boehnhard, D. P. Cruikshank, and A. Morbidelli, with the assistance of Renee Dotson. 2008.

Consolmagno, Guy. *Brother Astronomer: Adventures of a Vatican Scientist.* New York: McGraw-Hill, 2000.

Includes an account of Consolmagno's experiences on an Antarctic meteorite expedition. The Web pages for the ANSMET expeditions can be found at http://geology.cwru.edu/~ansmet.

Davies, John. *Beyond Pluto: Exploring the Outer Limits of the Solar System.* Cambridge: Cambridge University Press, 2001.

Written by an expert astronomer, this volume covers transneptunian objects and centaurs in general.

Levy, David, and Stephen J. Edberg. *Observing Comets, Asteroids, Meteors, and the Zodiacal Light.* Practical Astronomy Handbooks. Cambridge: Cambridge University Press, 1994.

Co-written by one of the most famous astronomers of recent times, David Levy, this handbook was designed for those who wish to do observing themselves, with or without a telescope.

McSween, Harry Y. *Meteorites and Their Parent Planets.* 2nd ed. Cambridge: Cambridge University Press, 1999.

McSween, a veteran meteorite researcher, wrote this useful overview of the study of meteorites and what is learned from them.

Richardson, Derek, and Kevin Walsh. "Binary Minor Planets." *Annual Review of Earth and Planetary Science.* Palo Alto, CA: Annual Reviews, 2006.

An excellent overview of asteroidal satellites from a more technical angle.

Sagan, Carl, and Ann Druyan. *Comet.* New York: Random House, 1985.

The late Carl Sagan, one of the greatest popularizers of science in the twentieth century, turns his attention to the small bodies.

Spencer, John, and Jacqueline Mitton, eds. *The Great Comet Crash: The Collision of Comet Shoemaker-Levy 9 and Jupiter.* Cambridge: Cambridge University Press, 1995.

David Levy's collaborators Carolyn and Gene Shoemaker contributed the foreword to this book about an object they discovered together, Comet Shoemaker-Levy 9, which later impacted Jupiter.

Stern, Alan, and Jacqueline Mitton. *Pluto and Charon: Ice Worlds on the Ragged Edge of the Solar System.* New York: John Wiley and Sons, 1997.

Alan Stern is the principal investigator of the *New Horizons* mission to Pluto.

Warner, Brian, and Alan W. Harris. *A Practical Guide to Lightcurve Photometry and Analysis.* New York: Springer-Verlag, 2006.

One of the most prolific lightcurve observers, Brian Warner, has coauthored this in-depth and technical treatment of the subject.

JOURNAL ARTICLES

Every year, dozens of articles about the newest research on asteroids, comets, transneptunian objects, and dwarf planets are written in scientific journals. The journals in which these articles are typically published include *Icarus, Meteoritics and Planetary Science, Astronomy and Astrophysics,* and *The Astronomical Journal,* among others. Lightcurve observations are often made by astronomy hobbyists, and their work is often published in the *Minor Planet Bulletin.* Check recent issues of these publications for articles on relevant topics. University and college libraries often have subscriptions to the print or electronic versions of these journals, and some of them also provide free access to older articles.

WEB SITES

American Museum of Natural History: http://www.amnh.org/exhibitions/permanent/meteorites/planets/crust.php.

This excellent set of online pages about asteroidal and planetary interiors and differentiation is hosted by the American Museum of Natural History. A popular-level description of the interior of Ceres is produced at *Astronomy* magazine's Web site at http://www.astronomy.com/asy/default.aspx?c=a&id=3478.

Antarctic Search for Meteorites (ANSMET) Program: http://geology.cwru.edu/~ansmet.

ANSMET is responsible for finding the vast majority of meteorites in the world's collections. Their homepage includes public information as well as advice for those researchers chosen to join the hunt. The team occasionally has been able to maintain Internet connections to the outside world and blog their adventures. Blog entries from seasons up through at least 2008–2009 can be found at http://www.humanedgetech.com/expedition/ansmet3/.

Asteroidal Satellite Discovery: http://www.boulder.swri.edu/~merline/press_release.

A description of the first ground-based detection of an asteroidal satellite by the discovery team, along with movies and images.

Asteroids: http://www.astro.cornell.edu/~richardson/asterseismo.html.

A technical description of seismic shaking on asteroids, including a set of explanatory images from *NEAR Shoemaker* and other missions.

"The Basics of Light": http://fuse.pha.jhu.edu/~wpb/spectroscopy/basics.html.

Contains additional detailed information on light.

Comets as Portents: http://www.bbc.co.uk/dna/h2g2/A3086101.

A recounting of how comets were sent as ill portents.

Cometary Studies: http://www.vigyanprasar.gov.in/dream/mar2001/comets.htm.

A more general history of cometary studies.

Damocloids: http://www.physics.ucf.edu/~yfernandez/lowtj.html.

A list of Damocloids and technical papers about objects that straddle the line between asteroids and comets are found here at Dr. Yan Fernandez' Web page.

Demotion of Ceres, Pallas, Juno, and Vesta from planet to asteroid: http://aa.usno.navy.mil/faq/docs/minorplanets.php.

Geminid Meteor Shower: http://www.earthsky.org/radioshows/51876/geminid-meteor-storm-from-3200-phaethon.

A discussion of the Geminid meteor shower and Phaethon, its parent body, were featured on the radio program *Earth and Sky* in 2007. A transcript and audio file can be found at this site.

History of the Status of Pluto: http://www.cfa.harvard.edu/icq/ICQPluto.html.

Hohmann Transfer Calculator: http://home.att.net/~ntdoug/smplhmn.html.

This online Hohmann Transfer calculator enables the user to input various starting and ending locations.

Hubble Space Telescope: http://hubblesite.org.

A great deal of data from the Hubble Space Telescope has been used to study protoplanetary disks around other stars. This Web site contains many of these images as well as explanations for them in their gallery.

IAU: Video of IAU Debate: http://www.astronomy2006.com/media-stream-archive.php.

IAU: Commentary on the IAU debate from a dissenting astronomer: http://www.lowell.edu/users/buie/pluto/iauresponse.html.

Irregular Satellites: http://www.dtm.ciw.edu/sheppard/satellites.

Atrove of information describing the irregular satellites of the giant planets put online by Scott Sheppard, who has discovered many of these objects.

Main-Belt Comets: http://star.pst.qub.ac.uk/~hhh/mbcs.shtml.

The main-belt comets (or activated asteroids) were discovered by Henry Hsieh, who provides more information, including links to more technical information, at this Web site. Further information on the MBCs can be found at http://www.planetary.org/blog/article/00000551/.

Mercury: http://www.lpl.arizona.edu/~sprague/planatmos/mercatmos.html.

Dr. Anne Sprague maintains a set of information about the atmosphere of Mercury at this Web site. Although it is not an asteroid, Mercury's atmosphere is similar in character to what one might hypothetically expect on Vesta, if conditions were right.

MESSENGER Mission: http://btc.montana.edu/messenger/instruments/grns.php and http://btc.montana.edu/messenger/instruments/xrs.php.

The MESSENGER mission to Mercury carries two instruments commonly found on small bodies missions: a gamma-ray spectrometer and an x-ray spectrometer. Popular-level descriptions and explanations of these instruments are found at these two Web sites.

Meteorites: University of Arizona: http://meteorites.lpl.arizona.edu.

Offers an abundance of information on meteorites.

Meteorites: Washington University: http://meteorites.wustl.edu.

More detailed material about meteorites.

Minor Planet Center: http://www.cfa.harvard.edu/iau/mpc.html.

Collects a large amount of data on near-earth objects, but is also designed to help observers report and catalog possible new discoveries.

Minor Planet Center: http://www.cfa.harvard.edu/iau/NEO/TheNEOPage.html.

The Minor Planet Center is the central clearinghouse for collecting NEO discoveries and observations and calculating orbits. Information about these observations, including technical information about astronomers, is found at this site.

Mission Descriptions: http://robotexplorers.blogspot.com.

Missions of all kinds, from those that flew to those that were more flights of fancy, are described at this blog, including small bodies missions as well as those with other targets.

NASA: Ames Research Center http://impact.arc.nasa.gov.

Offers additional NEO and impact information, including the Spaceguard report that helped begin the current age of asteroid surveys.

NASA: Comet Shoemaker-Levy 9: http://www2.jpl.nasa.gov/sl9/sl9.html.

The breakup of Comet Shoemaker-Levy 9 led to a large amount of material being posted on the then-new World Wide Web. This main page carries updates from a wide variety of sources. A pre-impact summary of what was expected from the Shoemaker-Levy impact can be found at http://www2.jpl.nasa.gov/sl9/background.html. Other comet impacts, these into the Sun, are described on http://science.nasa.gov/headlines/y2000/ast10feb_1.htm, along with images of them from the SOHO spacecraft. Another recent cometary breakup, that of Comet Schwassmann-Wachmann 3, was observed by the Hubble Space Telescope, among many other observatories. The HST images and descriptions can be seen at http://hubblesite.org/newscenter/archive/releases/2006/18.

NASA: Dawn Mission: http://dawn.jpl.nasa.gov.

Has a set of links about Vesta and its interior.

NASA: Electromagnetic Spectrum: http://imagine.gsfc.nasa.gov/docs/science/know_l1/emspectrum.html.

A general introduction to the electromagnetic spectrum and its different regions.

NASA: Genesis Mission: http://genesismission.jpl.nasa.gov/gm2/news/features/isotopes.htm.

Includes discussion of isotopic studies.

NASA: History of Asteroid Studies: http://dawn.jpl.nasa.gov/DawnCommunity/flashbacks/fb_06.asp.

NASA: Jet Propulsion Laboratory: http://neo.jpl.nasa.gov.

The Jet Propulsion Laboratory (JPL) is in charge of NASA's effort to track and characterize NEOs. This JPL site has a wide array of information, including the text of a recent report to the United States Congress compiled by experts and found at http://neo.jpl.nasa.gov/neo/report2007.html.

NASA: JPL Radar Research: http://echo.jpl.nasa.gov.

The JPL radar research page contains a large amount of information on asteroid observations, including both more technical and more popular treatments, and a large amount of images, movies, and links.

NASA: List of Planetary Missions: http://nssdc.gsfc.nasa.gov/planetary/projects.html.

This list of all NASA planetary missions offers links to NASA asteroid and comet missions, on which sites additional official information can be found. Most ongoing NASA missions have extensive education and public outreach (E/PO) activities, which also have a web presence. The general NASA E/PO site for planetary science—http://solarsystem.nasa.gov/planets/index.cfm—includes links to the specific types of objects described in this work.

NASA: Near-Earth Object Program: http://neo.jpl.nasa.gov.

Collects a large amount of news and general information on near-earth objects.

NASA: Page Defining Planets and Dwarf Planets: http://solarsystem.nasa.gov/planets/profile.cfm?Object=Dwarf&Display=OverviewLong.

NASA: Phoebe: http://saturn.jpl.nasa.gov/science/moons/moonDetails.cfm? page ID=12.

The official NASA Web page for Phoebe, satellite of Saturn and possible captured TNO.

NASA: Positions and Orbits: http://ssd.jpl.nasa.gov/?orbits.

The positions and orbits of comets, asteroids, and dwarf planets are provided by the NASA Jet Propulsion Laboratory. Views are available of the orbits of individual objects, or positions of small bodies in large regions of the solar system.

NASA: Remote Sensing Data: http://pdssbn.astro.umd.edu.

NASA maintains this central clearinghouse for a wide variety of remote sensing data from both missions and Earth-based research.

NASA: Sedna Discovery: http://www.nasa.gov/vision/universe/solarsystem/planet_like_body.html.

The discovery of Sedna, thought to be the only Oort cloud object currently known, is detailed at this site, with information about further Hubble Space Telescope observations available at http://hubblesite.org/newscenter/archive/releases/2004/14.

Northern Arizona University: http://www4.nau.edu/meteorite/Meteorite/Book-Minerals.html.

This site provides further information about the minerals found in meteorites, including chemical formulas and appearance.

Orbit Simulator: http://phet.colorado.edu/sims/my-solar-system/my-solar-system.swf.

This flash-based online orbit simulator includes several preset situations (including a Trojan asteroid-like case) and the ability to input positions and velocities to allow further exploration. A more-technical introduction to the mathematics behind orbital calculations, including several further references, is provided at http://mysite.du.edu/~jcalvert/phys/orbits.htm.

Planetary Society: http://www.planetary.org/explore/topics/near_earth_objects.

The Planetary Society offers this set of popular-level links on issues of near-Earth objects and the impact hazard they pose.

Pluto: http://www.johnstonsarchive.net/astro/pluto.html.

Offers a more-detailed discussion of Pluto's atmosphere, touching on other TNO atmospheres and with sections focusing on small body satellites.

Scientific American article about the "Planet Debate": http://www.sciam.com/article.cfm?chanID=sa006&articleID=93385350-E7F2-99DF-3FD6272BB4959038&pageNumber=2&catID=2.

Sky & Telescope magazine: http://skytonight.com/observing/objects/meteors/3304116.html.

A history of the scientific study of meteors.

Small Body Missions, non-U.S.: http://www.spacetoday.org/Japan/Japan/MUSES_C.html and http://www.esa.int/SPECIALS/Rosetta/index.html.

Most current-day missions have a presence on the Internet. Two non-U.S. small bodies missions with Web sites are Japan's Hayabusa and Europe's Rosetta.

Small Main Belt Asteroid Spectral Survey (SMASS): http://smass.mit.edu.
An excellent resource for asteroid spectra.

Stardust Mission: http://stardustathome.ssl.berkeley.edu.
This site is for those interested in taking part in the *Stardust* sample analysis and for obtaining information about the *Stardust* mission.

Tunguska Event: http://www.psi.edu/projects/siberia/siberia.html.
A different look at the impact hazard is provided by scientist and artist Bill Hartmann at this site, which details his attempt to make a painting of the Tunguska Event as scientifically accurate as possible.

United States Geological Survey: http://speclab.cr.usgs.gov/PAPERS.refl-mrs/refl4.html.
This USGS site has a detailed explanation of a wide variety of spectroscopic techniques at a higher technical level than presented here. The USGS also provides a library of reflectance spectra for minerals at http://speclab.cr.usgs.gov/spectral-lib.html.

University of Hawaii/David Jewett: http://www.ifa.hawaii.edu/faculty/jewitt/rubble.html. (http://www.ifa.hawaii.edu/~jewitt/coma.html) (http://www.ifa.hawaii.edu/~jewitt/tail.html)
The co-discoverer of the first Kuiper belt object, Dr. Dave Jewitt of the University of Hawaii, has this set of Web pages with information about comets and dwarf planets. A simplified explanation of how lag deposits form on comets is presented at the first site listed in this section. The latter two center on the coma and tails of comets.

"Windows to the Universe": http://www.windows.ucar.edu.
The "Windows to the Universe" Web site includes a set of Web pages about solar system formation at various levels of detail (in Spanish as well as English), and includes a reading list that offers books on solar system formation.

Index

About the Author

ANDREW S. RIVKIN was born in New York in 1969. As a boy, the *Viking* and *Voyager* missions led to an interest in astronomy that lasts to this day. A graduate of MIT with a doctorate in planetary sciences from the University of Arizona, Rivkin now works at the Johns Hopkins University Applied Physics Laboratory, specializing in observing asteroids and analyzing their compositions. When not at the telescope or studying data, he's likely at a baseball game or listening to the Beatles.